A Primer on Fluid Mechanics with Applications

Sudhir Ranjan Jain • Bhooshan S. Paradkar •
Shashikumar M. Chitre

A Primer on Fluid
Mechanics
with Applications

Springer

Sudhir Ranjan Jain
Theoretical Nuclear Physics and Quantum
Computing Section, Nuclear Physics
Division
Bhabha Atomic Research Centre
Mumbai, India

Bhooshan S. Paradkar
Department of Physics
Center for Excellence in Basic Sciences
Mumbai, India

Shashikumar M. Chitre
Department of Physics
(1936-2021) Prof. Chitre wrote this book
while at Center for Excellence in Basic
Sciences
Mumbai, India

ISBN 978-3-031-20486-9   ISBN 978-3-031-20487-6   (eBook)
https://doi.org/10.1007/978-3-031-20487-6

This Springer imprint is published by the registered company Springer Nature Switzerland AG
The registered company address is: Gewerbestrasse 11, 6330 Cham, Switzerland

To the memory of Professor S. M. Chitre
-BSP, SRJ

and

To my mother, Sudha Jain
-SRJ

Foreword

Fluid mechanics describes the collective motions of systems with a large number of components and provides essential descriptions of flow phenomena in a wide variety of systems of interest in science and engineering. Such phenomena include the motions of blood cells, flagella, and other components of biological organisms; the collective motion of particles in a gas or liquid; transport in plasmas ranging from the microscopic quark-gluon plasma in nuclei, to macroscopic plasmas of interest in nuclear fusion; and on an astronomical scale, the collective motions of matter and fields in stars and galaxies. As someone who has worked for many years on issues related to the statistical foundations of fluid mechanics, I greet the appearance of this new book by S. R. Jain, B. S. Paradkar, and S. M. Chitre with considerable pleasure and enthusiasm. It provides careful, clear, and elegant introductions to the basic principles and equations of fluid dynamics as well as discussions of many important applications and extensions of these equations to the description of fluid systems on all size scales. This book had its origins in the experiences of two of the authors, Jain and Paradkar, when serving as assistants to the late Prof. S. M. Chitre in India during some of the years that he taught a course in fluid mechanics to fourth-year physics students. Inspired by his teaching, they greatly extended and expanded his notes into this remarkable book. They honor his contributions and memory by including his name as a co-author of this book. Sudhir Jain is a very accomplished mathematical physicist with an interest in nonlinear dynamics and classical and quantum chaos theory among other topics. His co-author, Bhooshan Paradkar, is an engineering physicist, specializing in computational and theoretical plasma physics. The physical understanding and mathematical elegance of their research work and their teaching is displayed once again in their new book, *A Primer on Fluid Mechanics with Applications*. Here one finds very clear presentations of the mathematical and physical foundations of the equations of fluid mechanics as well as their applications to many scientific fields. One learns that the basic equations of fluid mechanics are obtained by combining mechanical conservation laws, typically of particles, momentum, and energy, with constitutive equations for quantities appearing in these equations such as the stress tensor or the energy current. Expressions for the constitutive quantities, such as Newton's law of friction or Fourier's law of heat conduction, allow for a closure of the equations where all quantities appearing in them are well defined. Moreover, the derivations are accompanied by very helpful analysis of the physical circumstances under which the equations are applicable. To be sure, the basic calculations of first course in fluid mechanics are given clear and careful treatments. These include, among others, derivations of the equations for incompressible and for compressible fluids, descriptions of Poiseuille flow and low Reynolds number flow around a sphere or an arbitrarily shaped object, and the use of complex variables to treat two-dimensional potential flows. A valuable chapter on the kinetic theory of gases clarifies for such systems the relation between the motions of the microscopic constituent particles of a gas and the macroscopic collective motion of the system as a whole. One can show that under certain conditions the flows appearing as steady solutions of the fluid equations are unstable under small perturbations. These hydrodynamic and related hydromagnetic instabilities are discussed in a separate chapter, where, among others, one finds treatments of the Rayleigh-Taylor and the Rayleigh-Bénard instabilities in fluids, and the Parker instability in an interstellar medium responsible for the observed lumpiness of interstellar media.

A particularly welcome and striking feature of this book is the presentation of a wide range of applications of fluid mechanics. Interesting techniques and applications often appear as sections in a chapter devoted to a more general topic. As examples, the chapter on low Reynolds number flows contains an example of the use of symmetry and time-reversal properties of the equations to draw some conclusions about the flagellar locomotion of micro-organisms. The chapter on shock waves contains a section on blast waves that is then applied to the waves produced in supernova explosions. In keeping with the principal aim of the book, separate chapters are devoted to applications of fluid mechanics to particular scientific areas. These include a chapter on physiological hydrodynamics including the analysis of blood flows in organs and arteries, a chapter on magnetohydrodynamics and plasma physics including a treatment of tearing instabilities in a plasma, a chapter on the tensor virial theorem including its application to the physics of neutron stars, and a chapter on water waves with a discussion of solitons and an instructive exercise devoted to drift waves produced by the wind over the water surface. Also included are a chapter on astrophysical applications with discussions of accretion disks and of the generation and maintenance of cosmic magnetic fields; a chapter on perhaps the most difficult fluid phenomena to explain quantitatively, namely, fluid turbulence; and a chapter on the behavior of quantum fluids with a discussion of quantized vortices and quantum turbulence.

I highly recommend that readers closely follow the details of the many applications of fluid mechanics throughout the book, for, taken together, they provide a valuable education on how to approach and treat complex physical phenomena using quantitative, analytical methods.

In summary, I am pleased to say that I find this book to be exceptionally interesting and worthwhile, and I recommend it to scientists and science students at all levels.

University of Maryland, College Park, MD, USA   J. Robert Dorfman
May 2022

Preface

Fluids surround us and influence our lives. Natural phenomena originating in fluids is part of our day-to-day experience. As common and accessible it is, it is much farther inaccessible in understanding. The main reason is the nonlinearity which is the essence of the subject. Fluid mechanics is the epitome of classical physics. All of classical "wisdom" is required to understand fluid behavior. It is also a fine illustration of a classical field theory. With these thoughts, the subject of Fluid Mechanics, in our opinion, forms an important component of any degree course in science and engineering.

The book has resulted from the course taught at the University of Mumbai-Department of Atomic Energy Centre for Excellence in Basic Sciences to students in the fourth year of the Master's degree course in Physics. The course was designed and taught for over ten years by late Professor Shashikumar M. Chitre and assisted by us. His handwritten notes form the content of many chapters of the book. We have tried to "complete the story," so to say, by adding all we could imagine and write. We can only hope that it meets Professor Chitre's vision.

Some two years ago, SRJ was invited to contribute to Springer briefs by a member of the Editorial Board, Professor B. Ananthanarayan. This book is a contribution in response to the invitation. Dr. Lisa Scalone, Editor, Springer took over after initial correspondence and has been advising us all along. Keeping the structure of the book unchanged, the suggestions received from her have been of great value in shaping the book to its present form. The book is meant for a one-semester course on Fluid mechanics where larger emphasis is to learn to apply the ideas to different areas of significance.

Mumbai, India
May 2022

Sudhir Ranjan Jain
Bhooshan S. Paradkar

Acknowledgments

Authors would like to record deep appreciation to Lisa Scalone for being a guide in our journey. SRJ, in particular, wishes to thank her for being so patient during the entire period which was strangely marred by continuous personal challenges, tragedies, and the loss of Professor Chitre (SMC). We thank B. Ananthnarayan, I.I.Sc., Bengaluru, for the initial invitation to SRJ. SRJ remembers the times he spent with SMC, getting initiated into the subject of hydrodynamic instabilities and other works by S. Chandrasekhar in the summer of 1984, and then, his companionship in teaching since 2010. Nonlinear plasma theory was introduced to SRJ by legendary plasma physicist, Late Professor P. K. Kaw. BSP would like to thank his teachers in the University of California San Diego, in particular to Professor Sergei Krasheninnikov, Professor Patrick Diamond, and Professor Eric Lauga for introducing him to the subject of fluid mechanics.

We are very grateful to Professor Jay Robert Dorfman for writing the Foreword. Having learned a lot from his beautiful books and research works, the authors are deeply privileged to have his words here.

SRJ thanks the Directors of Bhabha Atomic Research Centre (BARC) and UM-DAE Centre for Excellence in Basic Sciences during the last ten years for providing him the opportunity to contribute at many levels to the genesis of the centre. BSP also would like to thank his colleagues in the UM-DAE Centre for Excellence in Basic Sciences for their constant encouragement.

SRJ takes a great pleasure to thank Phalguni Shah, an outstanding former student of UM-DAE CEBS who wrote notes from the course in 2014 and made her meticulous work available to SRJ. In fact, we thank all the students over these years for their immense contributions by their strong participation. The authors thank another incredibly talented former student from UM-DAE CEBS, Amit Seta for drawing a detailed critique of the chapter on astrophysical fluid mechanics. We are indebted to Abhijit Bhattacharyya, who wrote Sect. 13.5 on the connection between black hole physics and fluid flow. SRJ thanks the "Class of 2010" students from Homi Bhabha National Institute and BARC where he taught the course "Nonlinear Plasma Theory" to exceptionally motivated people. We thank Komal Kumari and Garima Rajpoot for working out Chaps. 14 and 15, respectively, and providing their critique. We thank Komal for providing the picture of Lorenz attractor.

SRJ has been able to complete this book during the most difficult and trying period of life *only* because of enormous support of his wife, Alka, and encouragement from children, Manan and Swareena. Any expression in words would fall short.

Contents

Introduction

1

I did not say a word.
It was the bird that sang unseen
from the thicket.
The mango tree was shedding
its flowers upon the village road,
and the bees came humming one by one.

—*Rabindranath Tagore*

The constituents of our Universe may be described in the framework of the dynamics of fluids which refer to both gases and liquids. The fluid description is, of course, tenable provided a group of particles stay together for a sufficiently long duration. This may formally be characterized by the requirement that the mean free path of particles [26] should be small compared to characteristic length-scale of the system. There is an exception, however, when the fluid description is still valid even when the mean free path exceeds scale-length of the system. This arises when the magnetic field is present in an electrically conducting system and then charges spiral around the electromagnetic field lines and may be conceived as moving together perpendicular to the magnetic lines of force [107]. In such a situation the fluid description is valid provided the gyro-radius is small compared to scale-length of the system.

The main feature characterizing a fluid is that it can be deformed, causing a permanent rearrangement of the system, as distinct from a solid. The principal properties of a fluid are its inertia attributable to its mass, compressibility resulting from change in its density on application of pressure, and, viscosity which enables it to offer resistance to any shear applied to the system.

Continuum Hypothesis
Throughout this study it will be assumed that the fluid's macroscopic behaviour is such that it is continuous in structure, characterized by physical entities such as

mass, momentum, density, pressure, temperature, entropy, etc. associated with a small fluid element, which remain sensibly uniform over a typical volume element, $d\tau$. The fluid may be compressible or incompressible in character.

The continuum hypothesis enables us to adopt the simple concept of local velocity of a fluid element with the whole flow-field specified as an aggregate of such local velocities. If the fluid flow is denoted by the velocity at any point $\mathbf{r}$ at any time, t by $\mathbf{u}(\mathbf{r}, t)$ (in the Cartesian coordinate system, $\mathbf{r} = (x_1, x_2, x_3)$, with $\mathbf{u} = (u_1, u_2, u_3)$, then in the Eulerian picture [28], the fluid velocity $\mathbf{u}$ is specified instantaneously as a function of position, $\mathbf{r}$ and time, t referred to some fixed axes. Thus, the Eulerian system may be thought of providing an instantaneous picture (snapshot) of the spatial distribution of fluid velocity and other quantities such as pressure, density, entropy, etc.

The Lagrangian picture, on the other hand, specifies the velocity $\mathbf{u}$ at time t, of any selected fluid element described by its initial position, $\mathbf{r}_0$ at the initial instant, t_0, by $\mathbf{u}((\mathbf{r}_0, t_0), \mathbf{r}, t)$. The Lagrangian method thus tracks the dynamical history of a selected fluid element. Such a prescription makes use of the fact that some of the physical entities refer not merely to certain positions in space, but also to the initial conditions of the fluid element. The operator, $d/dt = \partial/\partial t + v.\nabla$ describes the rate of change of any fluid quantity resulting from time rate of change together with the change arising from the convective motion of the fluid element. In our analysis, the velocity field $\mathbf{u}(\mathbf{r}, t)$ will be the primary variable to characterize the motion and other flow quantities such as pressure, density, entropy, etc. will likewise be functions of $\mathbf{r}$ and t. The motion will be described steady when the velocity $\mathbf{u}$ is independent of time, but has only spatial dependence.

Acceleration of a Fluid Element

Suppose a fluid element which is at position, $\mathbf{r}$ at any instant, t moves to a nearby position $\mathbf{r} + \delta\mathbf{r}$ at time, $t + \delta t$ where $\delta\mathbf{r} = \mathbf{u}\delta t$, then the velocity change in small time interval δt is

$$\mathbf{u}(\mathbf{r} + \delta\mathbf{r}, t + \delta t) - \mathbf{u}(\mathbf{r}, t) = \mathbf{u}(\mathbf{r}, t) + \delta t \frac{\partial \mathbf{u}}{\partial t} + (\delta\mathbf{r}.\nabla)\mathbf{u} + O(\delta t^2). \tag{1.1}$$

Hence, the acceleration of fluid element at $(\mathbf{r}, t)$ is given by

$$\frac{d\mathbf{u}}{dt} = \frac{\partial \mathbf{u}}{\partial t} + \left(\frac{\partial \mathbf{r}}{\partial t}.\nabla\right)\mathbf{u}. \tag{1.2}$$

Here, $\frac{\partial \mathbf{u}}{\partial t}$ expresses the local time rate of the change of the velocity, while $(\mathbf{u}.\nabla)\mathbf{u}$ represents the convective change resulting from the transport of the material element to a different position. In fact, the operator, $d/dt = \partial/\partial t + \mathbf{u}.\nabla$ may be regarded as describing the rate of change for any physical quantity associated with the fluid element such as the velocity, pressure, density, etc., resulting from a local time rate of change together with the change in the entity arising from the convective motion on account of advection. Thus,

$$\frac{d\boldsymbol{u}}{dt} = \frac{\partial \boldsymbol{u}}{\partial t} + (\boldsymbol{u}.\nabla)\boldsymbol{u},$$

$$\frac{d\rho}{dt} = \frac{\partial \rho}{\partial t} + (\boldsymbol{u}.\nabla)\rho,$$

$$\frac{dp}{dt} = \frac{\partial p}{\partial t} + (\boldsymbol{u}.\nabla)p. \tag{1.3}$$

1.1 Hydrodynamic Description: Validity

If the collection of particles remains together for a sufficiently long period of time, we may describe it hydrodynamically. For this to happen the mean free path, λ of the particles must be small compared to the characteristic length-scale, L of the system ($\lambda \ll L$).

The collective behaviour is also characterized by the length-scales over which pressure, P, density, ρ, and temperature, T vary. These are termed as "scale-heights". These are defined as follows:

$$(\text{Density scale-height}), \ H_\rho = \frac{dz}{d(\log_e \rho)} = \frac{dz}{d\rho/\rho},$$

$$(\text{Pressure scale-height}), \ H_P = \frac{dz}{d(\log_e P)} = \frac{dz}{dP/P},$$

$$(\text{Temperature scale-height}), \ H_T = \frac{dz}{d(\log_e T)} = \frac{dz}{dT/T}. \tag{1.4}$$

Mean molecular weight is another variable to which we may assign a length-scale.

For neutral atoms/molecules, the mean free path is given by $\lambda = (n\sigma)^{-1}$ where n is the number density and σ is the collision cross-section. In the case of a plasma, consisting of ions and electrons, the mean free time in which an ion with a mean velocity ($\sim$ thermal velocity, V_{th}) undergoes a cumulative deflection due to Coulomb scattering from the constituents of the ionized gas with density, n_i of ions per unit volume is roughly

$$t_{\text{def}}^{(i)} \sim 3 \times 10^{-12} \frac{V_{th}^3}{n_i \log_e \Lambda} \text{s}, \quad \Lambda = 1.24 \times 10^4 \frac{T^{3/2}}{n_i^{1/2}},$$

$$\lambda \sim V_{th} t_{\text{def}} = 3 \times 10^{-12} \frac{V_{th}^4}{n_i \log_e \Lambda} \sim \frac{T^2}{n_i} \tag{1.5}$$

Due to weak dependence on Λ, quite expectedly, large mean free path is equivalent to higher temperature or a rarefied medium.

Exercise 1.1 The number density of air in the Earth's atmosphere drops from $2.5 \times 10^{19}\,\text{cm}^{-3}$ with a scale-height of roughly $8.5\,\text{km}$. Find the height beyond which continuum approximation is not tenable. Assume constant collision cross-section ($\sigma = 10^{-15}\,\text{cm}^{-2}$) and scale-height throughout the atmosphere.

1.1.1 Streamlines, Streaklines, Pathlines

There are various distinct and interesting ways of visualizing the fluid flow, we discuss these here.

A line in the fluid whose tangent is everywhere parallel to the velocity vector, $\mathbf{u} = (u_1, u_2, u_3)$ instantaneously is called a *streamline*. At any instant, the family of streamlines are solutions of

$$\frac{dx_1}{u_1(\mathbf{r}, t)} = \frac{dx_2}{u_2(\mathbf{r}, t)} = \frac{dx_3}{u_3(\mathbf{r}, t)} = ds. \qquad (1.6)$$

For instance, if $\mathbf{u} = (x_2, x_1, 0)$, we have for streamlines,

$$\frac{dx_1}{x_2} = \frac{dx_2}{x_1}. \qquad (1.7)$$

The streamlines are given by $x_1^2 = x_2^2 + c$ where c is a constant.

A *pathline* is a curve followed in time by a fluid particle as it flows. The particle is moving with the fluid at its local velocity. Thus, pathlines must satisfy the equations:

$$\frac{dx_i}{dt} = u_i(x_i, t). \qquad (1.8)$$

The equation of the pathline that passes through the point (x_0, y_0, z_0) at time $t = 0$ will be solution of (1.8) with initial condition at $t = 0$ where $x = x_0, y = y_0, z = z_0$. The solution will therefore yield a set of equations of the form, $x_i = x_i(x, y, z, t)$.

Streakline is a curve traced out by a neutrally buoyant marker fluid that is continuously injected into a flow field at a fixed point in space. The marker may be dye or smoke.

Let us illustrate the difference between the three representations by the aid of an example. Consider a two-dimensional flow field defined by $u_1 = x_1(1+t)$, $u_2 = x_2$, $u_3 = 0$. Using (1.6), we have

$$\frac{dx_1}{ds} = x_1(1+t), \text{ implying } x_1 = c_1 e^{(1+t)s},$$

$$\frac{dx_2}{ds} = x_2, \text{ implying } x_2 = c_2 e^{s}. \tag{1.9}$$

These are parametric equations of the streamlines in the $x_1 x_2$-plane. If the streamlines pass through a point, $(1, 2)$, then we may use the initial conditions that when $s = 0$, $x_1 = 1$, $x_2 = 2$, thus $c_1 = 1$, $c_2 = 2$. The solution for the streamlines is then $x_1 = e^{(1+t)s}$, $x_2 = 2e^{s}$. Note that streamlines change with time. Suppose the streamline passes through $(1, 2)$ at time $t = 0$, then $x_1 = e^{s}$, $x_2 = 2e^{s}$. The equation for streamline is then $x_2 = 2x_1$.

For finding the pathline, we need to solve the equations:

$$\frac{dx_1}{dt} = x_1(1+t), \quad \frac{dx_2}{dt} = x_2. \tag{1.10}$$

The solutions are simple: $x_1 = c_1 e^{t(1+t/2)}$, $x_2 = c_2 e^{t}$. As considered for streamlines, let us assume that path of the particle passes through $(1, 2)$ at $t = 0$, Then, $c_1 = 1$, $c_2 = 2$, entailing $x_1 = e^{t(1+t/2)}$, $x_2 = 2e^{t}$. Eliminating t from these we obtain the equation for the flow field

$$x_1 = \left(\frac{x_2}{2}\right)^{\log(e\sqrt{x_2/2})}. \tag{1.11}$$

Finally, for the streakline, the equations are the same as for path line and the solutions are the same. However, here we need to consider the condition that $(x_1, x_2) = (1, 2)$ at a time, τ. This implies that $c_1 = e^{-\tau(1+\tau/2))}$ and $c_2 = 2e^{-\tau}$. Thus,

$$x_1 = e^{t(1+t/2)-\tau(1+\tau/2)}, \quad x_2 = 2e^{t-\tau}. \tag{1.12}$$

These are parametric equations for the streak lines that pass through $(1, 2)$; they are valid for all times. If we choose $t = 0$, and eliminate τ, then the equation for streamline is

$$x_1 = \left(\frac{x_2}{2}\right)^{\log(e\sqrt{2/x_2})}. \tag{1.13}$$

Figure 1.1 illustrates how these three representations are so different for the same fluid flow. However, in the neighbourhood of the point $(2, 1)$, all of them agree.

Fig. 1.1 Streamline (blue), pathline (orange), and streakline (green) for a fluid flow (see details in the text)—they are quite different

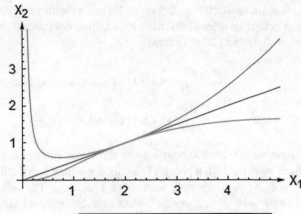

Fig. 1.2 Water in the container flowing out through a hole provides an illustration of Torricelli's law (See text)

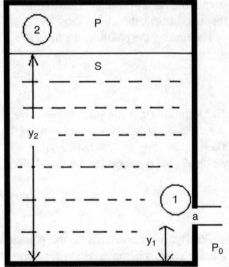

1.1.2 Torricelli's Law

One of the earliest introductory ideas pertaining to fluid flows result by our asking the flow of liquid out of a tank through a hole. Assuming the tank is filled with a liquid of density, ρ, with the cross-section of the hole, a much smaller than the cross-section of the cylinder, S, we wish to determine how the speed of the liquid leaving the hole depends on the distance h above the hole. The assumption, $a \ll S$ helps in assuming that the liquid at the top of the tank, where the pressure is P, is at rest approximately. If we apply Bernoulli's theorem at two points, '1' and '2', we have (Fig. 1.2)

$$P_{\text{atm}} + \frac{1}{2}\rho v_1^2 + \rho g y_1 = P + \rho g y_2, \tag{1.14}$$

Fig. 1.3 Water particles
coming out of an orifice in the
cylinder kept on a stool,
follow a parabolic path.
However, what we see at a
given instant is a streakline
which is not a parabola

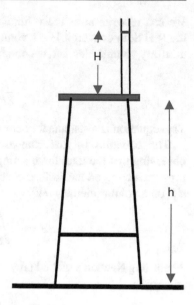

Denoting $y_2 - y_1$ by h, v_1 is given by

$$v_1 = \sqrt{\frac{2(P - P_{\text{atm}})}{\rho} + 2gh}. \tag{1.15}$$

Exercise 1.2 Figure 1.3 shows a commonplace instance where water is coming out
of a tank placed on a stool. The height of tank and stool are respectively H and h.
The cross-sectional area of the cylindrical tank is S and cross-section of the orifice
from where water comes out is a. Pressure at the top of the tank and sides of the jet
is assumed to be atmospheric pressure. Find the equation of the jet we observe.

1.2   Conservation Laws and Governing Equations

1.2.1   Continuity Equation

Consider a fluid with mass density ρ moving with a local velocity $\mathbf{v}$ under the
influence of pressure, $p(x, y, z, t)$. From a given volume, $\mathcal{V}$ enclosed by a surface,
S, transit of an element of this fluid from this will result in loss of mass, m which is
exactly equal to the fluid crossed through the area element. Thus,

$$\frac{\partial m}{\partial t} = -\int \rho \mathbf{v}.\hat{n}dS. \tag{1.16}$$

We can re-write m as the volume integral of density, and, use Green theorem on the R.H.S. to convert it into a volume integral. Since the equation will hold for any arbitrary volume element, we obtain the continuity equation:

$$\frac{\partial \rho}{\partial t} + \nabla.(\rho \mathbf{v}) = 0. \tag{1.17}$$

This equation is a statement of *conservation of mass*.

The equation for the change of velocity, $\mathbf{v}(x, y, z, t)$ can be obtained by observing that the total force acting on the volume $\mathcal{V}$ is the surface integral of the pressure acting on the surface $\mathcal{S}$ enclosing the volume. That is, fluid exerts a force dF on a volume element dV:

$$d\mathbf{F} = -\int_S p\hat{n}dS = -\int \nabla p dV. \tag{1.18}$$

According Newton's second law,

$$\rho \frac{d\mathbf{v}}{dt} = -\nabla p. \tag{1.19}$$

Clearly, as

$$d\mathbf{v} = \frac{\partial \mathbf{v}}{\partial t}dt + \frac{\partial \mathbf{v}}{\partial x}dx + \frac{\partial \mathbf{v}}{\partial y}dy + \frac{\partial \mathbf{v}}{\partial z}dz, \tag{1.20}$$

we have the convective derivative of velocity field, (1.18). With this, we arrive at the well-known result of Euler [28]:

$$\frac{\partial \mathbf{v}}{\partial t} + (\mathbf{v}.\nabla)\mathbf{v} = -\frac{1}{\rho}\nabla p + \mathbf{f} \tag{1.21}$$

where $\mathbf{f}$ is an external force. Equation (1.21) can be re-written in terms of vorticity, $\boldsymbol{\omega} = \nabla \times \mathbf{v}$:

$$\frac{\partial \mathbf{v}}{\partial t} + \frac{1}{2}\nabla \mathbf{v}^2 - \mathbf{v} \times \boldsymbol{\omega} = -\frac{1}{\rho}\nabla p + \mathbf{f}. \tag{1.22}$$

We have used the identity,

$$\frac{1}{2}\nabla \mathbf{v}^2 = \mathbf{v} \times (\nabla \times \mathbf{v}) + (\mathbf{v}.\nabla)\mathbf{v}. \tag{1.23}$$

As in continuity equation, denoting entropy density by s, its change is governed by

$$\frac{\partial(\rho s)}{\partial t} + \nabla.(\rho s \mathbf{v}) = 0. \tag{1.24}$$

1.2.2 Boussinesq Approximation

Boussinesq was the first to note several situations of practical importance where the density change results largely due to variation in temperature and there is negligible change on account of pressure variations. The origin of simplification may be attributed to the smallness of coefficient of volume expansion, α defined as

$$\alpha = -\frac{1}{\rho}\left(\frac{\partial \rho}{\partial T}\right) = \frac{1}{V}\left(\frac{\partial V}{\partial T}\right) \simeq O\left(\frac{1}{T}\right) \tag{1.25}$$

where $V = 1/\rho$ is the specific volume. However, there is one important exception, namely the density variation associated with the gravitational force, $(\delta\rho)\mathbf{g}$ in the equation of motion, which cannot be ignored because the resulting gravitational acceleration could be quite large. Accordingly, in the Boussinesq approximation the density, ρ is treated to be uniform in all the terms of equation of motion, except the one associated with gravity. The mass conservation equation then becomes

$$\nabla \cdot \mathbf{u} = 0 \tag{1.26}$$

1.2.3 Anelastic Approximation

Through the interior of a star the density varies by several orders of magnitude which makes the Boussinesq approximation unsuitable for application to stars. The anelastic approximation, on the other hand, expresses the density $\rho\,(\mathbf{r}, t)$ as

$$\rho\,(\mathbf{r}, t) = \overline{\rho}\,(\mathbf{r}, t) + \delta\rho\,(\mathbf{r}, t)$$

where $\overline{\rho}\,(\mathbf{r}, t)$ is the horizontal average and $\delta\rho$, the fluctuation density. The mass conservation equation then may be written as

$$\nabla \cdot (\overline{\rho}\mathbf{u}) = 0. \tag{1.27}$$

1.2.4 Validity of the Approximations

The fluid is designated to be incompressible if the density, ρ cannot be sensibly altered by application of pressure, though the change in density may result from thermal effects, i.e.

$$\delta\rho = \left(\frac{\partial\rho}{\partial p}\right)_T \delta p + \left(\frac{\partial\rho}{\partial T}\right)_p \delta T, \tag{1.28}$$

where the first term on the right hand side is deemed to be negligible compared to the second term.

Denoting the spatial distribution of velocity $\mathbf{u}$ and other flow quantities by a characteristic scale-length L and the velocity by U; the order of magnitude of the spatial derivatives of velocity, $\mathbf{u}$ is given by U/L. The velocity distribution is supposed to be solenoidal provided

$$|\nabla \cdot \mathbf{u}| = \left|\frac{1}{\rho}\frac{d\rho}{dt}\right| \ll U/L.$$

Let us consider an ideal isentropic flow in the absence of viscosity and thermal conductivity i.e.

$$dS/dt = 0 \quad \text{or} \quad dp/dt = C_s^2 (d\rho/dt). \tag{1.29}$$

Appealing to the equation of motion,

$$\rho\frac{d\mathbf{u}}{dt} = -\nabla p + \rho\mathbf{g}, \tag{1.30}$$

with $\mathbf{g}$ as the gravitational body force and then scalar multiplying by $\mathbf{u}$, we recover

$$\rho\frac{d}{dt}\left(\frac{1}{2}u^2\right) = -\mathbf{u} \cdot \nabla p + \rho\mathbf{u} \cdot \mathbf{g}. \tag{1.31}$$

Hence, the isentropic condition implies

$$\left|\frac{1}{\rho}\frac{d\rho}{dt}\right| = \left|\frac{1}{\rho C_s^2}\frac{dp}{dt}\right| = \left|\frac{1}{\rho C_s^2}\frac{\partial p}{\partial t} + \frac{\mathbf{u} \cdot \nabla p}{\rho C_s^2}\right|$$

which may be rewritten to

$$\left|\frac{1}{\rho C_s^2}\frac{\partial p}{\partial t} - \frac{\mathbf{u} \cdot \nabla\left(\frac{1}{2}u^2\right)}{C_s^2} + \frac{\mathbf{u} \cdot \mathbf{g}}{C_s^2}\right| \ll \frac{U}{L}$$

i.e.

$$\left|\frac{1}{\rho C_s^2}\frac{\partial p}{\partial t} - \frac{U^3}{LC_s^2} + \frac{gU}{C_s^2}\right| \ll \frac{U}{L} \tag{1.32}$$

requiring individual terms; $U^3/LC_s^2 \ll U/L$ and $gU/C_s^2 \ll U/L$ leading to

$$\frac{U^2}{C_s^2} \ll 1 \quad \text{and} \quad L \ll \frac{C_s^2}{g} \equiv \frac{p}{\rho g} = H_p. \tag{1.33}$$

For low frequency phenomena, $|\partial p/\partial t| \simeq \rho U L \omega^2$ the condition reduced to

$$\frac{\omega^2 L^2}{C_s^2} \ll 1 \tag{1.34}$$

If the representative frequency of the temporal variation is U/L then we recover the condition $U^2/C_s^2 \ll 1$. Thus, the incompressibility condition and the Boussinesq approximation holds for the typical velocities, U small compared to the sound speed C_s and typical length-scale, L small compared to the local scale-height, H_p [106].

1.3 Stress-Strain Relationship

There are two kinds of forces acting on the bulk of a fluid:

1. Long range forces such as gravity, electromagnetic and rotational forces which extend over whole of the interior of the body and may be considered to be uniform over a very small volume element of the fluid. These body forces, $\mathbf{F}(\mathbf{r}, t)$ equally on all the matter within a small volume element and $\mathbf{F}(\mathbf{r}, t)\,\rho d\tau$ may be regarded as the total body force acting on the fluid element, $d\tau$ at position $\mathbf{r}$ at time t.
2. Short range forces are of direct atomic and molecular origin and decrease rapidly with increasing distance between the interacting particles.

Considering a plane surface element in the fluid at position $\mathbf{r}$ with the associated unit normal, $\mathbf{n}$, the short range forces are the total force exerted on the fluid on one side of the element by the fluid on the other side. Then, the total surface force, $\mathbf{S}(\mathbf{n}, \mathbf{r}, t)\,\delta A$ will be proportional to the surface area δA and is designated the stress defined as the force per unit area, representing the surface forces due to fluid on the other side of an element of area, $\mathbf{n}\delta A$ at $(\mathbf{r}, t)$ acting on fluid on the other side with $\mathbf{S}(\mathbf{n}, \mathbf{r}, t)$ acting on the side of the fluid with the unit normal, $\mathbf{n}$ pointing away from the surface. For characterizing the force per unit area, namely the stress, consider a small tetrahedron OABC (Fig. 1.4) inscribed completely in the fluid [9] with OA, OB, OC along the three co-ordinate axes with O being the origin. Let δA_1, δA_2, δA_3 and δA be small planar areas of the sides OBC, OAC, OAB and ABC respectively (Fig. 1.4). Also, $\mathbf{a}, \mathbf{b}, \mathbf{c}$ be the unit vectors parallel to the axes OX_1, OX_2, OX_3 and $\mathbf{n}$ be the unit vector perpendicular to the plane ABC. Consider all the forces acting on the fluid

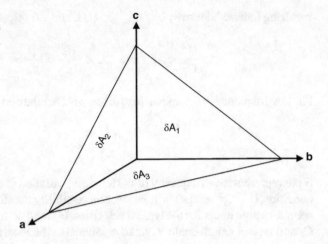

Fig. 1.4 A tetrahedron-shaped volume element. The three orthogonal faces are shown as $\delta A_1, \delta A_2, \delta A_3$

within the tetrahedron of small volume, $d\tau$ whose faces have axes $\delta A_1, \delta A_2, \delta A_3$ and δA. Then, the total force in the i-th direction is given by

$$\rho F_i d\tau + S_i\,(-\mathbf{a}, \mathbf{r}, t)\,\delta A_1 + S_i\,(-\mathbf{b}, \mathbf{r}, t)\,\delta A_2 + S_i\,(-\mathbf{c}, \mathbf{r}, t)\,\delta A_3 + S_i\,(\mathbf{n}, \mathbf{r}, t)\,\delta A$$

with $-\mathbf{a}, -\mathbf{b}, -\mathbf{c}$ being the unit outward normals. The equation of motion then becomes

mass × acceleration = resultant body force + surface forces

$$= \rho F_i d\tau + \delta A\,[-n_1 S_i\,(\mathbf{a}) - n_2 S_i\,(\mathbf{b}) - n_3 S_i\,(\mathbf{c}) + S_i\,(\mathbf{n})]$$

Dividing both sides by δA and letting $\delta\tau \to 0$, we obtain

$$S_i\,(\mathbf{n}) = n_1 S_i\,(\mathbf{a}) + n_2 S_i\,(\mathbf{b}) + n_3 S_i\,(\mathbf{c})$$

$$= n_j\,\left[a_j S_i\,(\mathbf{a}) + b_j S_i\,(\mathbf{b}) + c_j S_i\,(\mathbf{c})\right]$$

$$= n_j \sigma_{ij}$$

This establishes that the stresses for an arbitrary direction of a small element is related to stresses at the point $(\mathbf{r}, t)$ for any orthogonal surface elements. Then, the i-th component of the force exerted across the element; $\mathbf{n}\delta A = S_i\,(\mathbf{n})\,\delta A = n_j \sigma_{ij} \delta A$, where σ_{ij} is the i-th component of the force per unit area, i.e. the stress acting across a surface element, with normal in the j-th direction. The diagonal elements of σ_{ij} are designated normal stresses or pressure and off-diagonal terms the shearing stresses resulting from velocity gradients. Clearly, for a fluid at rest, only normal stresses or pressure will act, such fluids being normally in the state of compression. Thus, $\sigma_{ij} = -p\delta_{ij}$ for a fluid at rest is the isotropic tensor. This, stress-strain relationship is, of course, not valid when the fluid is in motion and it is then customary to decompose the stress tensor, σ_{ij} into an isotropic part, $-p\delta_{ij}$, along with a remaining non-isotropic part, to express σ_{ij} as $-p\delta_{ij} + f_{ij}$, with the isotropic part resulting from imposition of bulk compression and the non-isotropic part arising from the deformation of the fluid by motion. The latter part of stresses depend on the rate of strain tensor,

$$e_{ij} = \frac{1}{2}\left(\frac{\partial u_i}{\partial x_j} + \frac{\partial u_j}{\partial x_i}\right). \tag{1.35}$$

For analysing the motion in the neighbourhood of a point we consider a fluid velocity, $\mathbf{u}(\mathbf{r}, t)$ at a position $(\mathbf{r}, t)$ and simultaneous velocity at neighbouring $(\mathbf{r} + \delta\mathbf{r}, t)$ at the same instant is $\mathbf{u} + \delta\mathbf{u}$, where

$$\delta u_i = \partial x_j \frac{\partial u_i}{\partial x_j} = \partial x_j e_{ij} + \partial x_j \eta_{ij}, \tag{1.36}$$

with e_{ij} and η_{ij} are symmetric and anti-symmetric parts respectively of the second rank tensor, which may be represented as

$$\begin{aligned} e_{ij} &= \frac{1}{2}\left(\frac{\partial u_i}{\partial x_j} + \frac{\partial u_j}{\partial x_i}\right) \\ \eta_{ij} &= \frac{1}{2}\left(\frac{\partial u_i}{\partial x_j} - \frac{\partial u_j}{\partial x_i}\right) = -\frac{1}{2}\epsilon_{ijk}\omega_k \end{aligned} \tag{1.37}$$

where $\delta u_i = -\frac{1}{2}\epsilon_{ijk}\delta x_j\omega_k$ is the i-th component of the vector, $\frac{1}{2}\boldsymbol{\omega} \times \delta\mathbf{r}$ representing the velocity produced at position $\delta\mathbf{r}$ relative to position, $\mathbf{r}$, about which there is a rigid body rotation $\frac{1}{2}\boldsymbol{\omega}$. It is customary to represent $\boldsymbol{\omega} = \nabla \times \mathbf{u}$ as the local vorticity of the fluid motion, and express the velocity in the neighbourhood of a position $(\mathbf{r}, t)$ as

$$\begin{aligned} u_i(\mathbf{r} + \delta\mathbf{r}, t) &= u_i(\mathbf{r}) + \delta r_j e_{ij} + \delta r_j \eta_{ij} \\ &= u_i(\mathbf{r}) + \delta u_i(\mathbf{r}), \end{aligned}$$

here, $u_i(\mathbf{r})$ denotes uniform translation, $\delta r_j e_{ij}$ represents pure straining and $\delta r_j \eta_{ij}$ is a rigid rotation [9].

We are now in position to derive stress-strain relationship by noting that

$$\sigma_{ij} = -p\delta_{ij} + f_{ij} + \eta_b \Delta\delta_{ij}, \tag{1.38}$$

where, $\Delta = \frac{\partial u_i}{\partial x_i} = e_{ii}$ is the rate of dilation. Assuming the rate of deformation to be small, and only the first order terms in the expression of $f_{ij}(e_{ij})$ may be written as a series expansion in terms of e_{ij}, it is convenient to write

$$f_{ij} = A_{ij} + B_{ijkl}e_{kl} + O\left(e_{kl}^2\right), \tag{1.39}$$

with A_{ij}, B_{ijkl} being the constants characterizing the fluid. With the fluids having largely isotropic structure with no directional preference A_{ij} and B_{ijkl} to be isotropic tensor. Thus two tensors, A_{ij} and B_{ijkl} may be expressed as

$$A_{ij} = K\delta_{ij} \tag{1.40}$$

and

$$B_{ijkl} = \mu\delta_{ik}\delta_{jl} + \psi'\delta_{il}\delta_{jk} + \mu''\delta_{ij}\delta_{kl}. \tag{1.41}$$

Recalling that the stress tensor, σ_{ij} is symmetrical in i and j, so must B_{ijkl} in j and i be, thus implying $\mu = \mu'$. Using the definition $p = -\frac{1}{3}\sigma_{ij} + \eta_b\Delta\delta_{ii}$, we get $A_{ii} + B_{iikl}e_{kl} = 3\eta_b\Delta$, i.e.

$$k\delta_{ii} + 2\mu\delta_{ik}\delta_{il}e_{kl} + \mu''\delta_{ii}\delta_{kl}e_{kl} = 3\eta_b\Delta$$

or,

$$3K + \left(2\mu + 3\mu''\right)e_{kk} = 3\eta_b\Delta,$$

for all e_{kk}, which implies $K = \eta_b\Delta$ and $2\mu + 3\mu'' = 0$. This gives stress-strain relationship the form

$$\sigma_{ij} = -p\delta_{ij} + \mu_s\left(2e_{ij} - \frac{2}{3}\Delta\delta_{ij}\right) + \eta_b\Delta\delta_{ij}. \tag{1.42}$$

1.4 Conservation of Momentum

The Navier-Stokes equation of motion for a fluid in its most general form relates the rate of change of momentum of a fluid element and the sum of all the forces (body and surface stresses) acting on the fluid element with mass, $\rho d\tau$ occupying a small volume element $d\tau$ at a point $(\mathbf{r}, t)$. The momentum conservation equation may be expressed as

$$\text{mass} \times \text{acceleration} = \text{Body force} + \text{Surface force}.$$

For a body of fluid of volume τ enclosed by a surface S, the momentum $= \int_\tau \rho \mathbf{u} d\tau$ with its rate of change given by

$$\int_\tau \rho \frac{du_i}{dt} d\tau = \int_\tau \rho F_i d\tau + \int_S \sigma_{ij} n_j dS$$

$$= \int_\tau \rho F_i d\tau + \int_\tau \frac{\partial \sigma_{ij}}{\partial x_j} d\tau.$$

This holds for all points of the fluid occupying a volume, τ enclosed by surface, S, so that

$$\rho \frac{du_i}{dt} = \rho F_i + \frac{\partial \sigma_{ij}}{\partial x_j}$$

$$= \rho F_i - \frac{\partial p}{\partial x_i} + \mu_s \left(\frac{\partial^2 u_i}{\partial x_j \partial x_j} + \frac{1}{3} \frac{\partial \Delta}{\partial x_i} \right). \tag{1.43}$$

F_i is the i-th component of the body force and σ_{ij} is the surface stress $= -p\delta_{ij} + \mu_s \left(2e_{ij} - \frac{2}{3} \Delta \delta_{ij} \right) + \eta_b \Delta \delta_{ij}$. Here, μ_s and η_b are respectively the constants designating shear and bulk viscosities. For an incompressible fluid, $\Delta \equiv \nabla \cdot \mathbf{u} = \partial u_i / \partial x_i = 0$, the Navier-Stokes equation simplifies to

$$\rho \frac{d\mathbf{u}}{dt} = -\nabla p + \rho \mathbf{F} + \mu_s \nabla^2 \mathbf{u}, \tag{1.44}$$

$\mathbf{F}$ being the body force. Provided the body force, $\mathbf{F}$ and the shear viscosity are known, the equations of mass and momentum conservation along with the equation of state $p = p(\rho)$ determine the flow completely with the prescribed boundary conditions. This brings us to the thermodynamic characteristic of the fluid (e.g. entropy, temperature etc.).

1.5   Energy

Note the conservation of energy is the first integral of the momentum conservation equation. The rate of change of work is being done on the fluid contained in a volume τ enclosed by a surface S is the sum of $\int_\tau u_i F_i \rho d\tau$ resulting from the body forces along with $\int_S u_i \sigma_i j n_j dS = \int_\tau \frac{\partial}{\partial x_j} (u_i \sigma_{ij}) d\tau$ arising from surface stresses exerted at the boundary of the fluid volume by the surrounding medium. Hence the total rate of working per unit mass of the fluid is

$$u_i F_i + \frac{u_i}{\rho} \frac{\partial \sigma_{ij}}{\partial x_j} + \frac{\sigma_{ij}}{\rho} \frac{\partial u_i}{\partial x_j} = u_i \frac{du_i}{dt} + \frac{\sigma_{ij}}{\rho} \frac{\partial u_i}{\partial x_j} \tag{1.45}$$

using the momentum conservation equation, $\rho \frac{du_i}{dt} = \rho F_i + \frac{\partial \sigma_{ij}}{\partial x_j}$. There is thus a gain in kinetic energy $(= d\left(u_i^2/2\right)/dt)$ of bulk motion of the element resulting from the rate of working of the body force $\mathbf{F}$ and from one part of the rate of working of the surface forces, while the remaining surface term, $(\sigma_{ij}/rho)\,\partial u_i/\partial x_j$ is associated with the work done in the deformation of the fluid element without altering the velocity with time, which manifests entirely as an increase in the internal energy of the fluid. In addition, there will be heat flowing in or out of the fluid volume through either the process of conduction or radiation which may be expressed as heat gained by a fluid mass across the boundary surface, $S = \int_S \kappa \frac{\partial T}{\partial x_i} n_i dS = \int_\tau \frac{\partial}{\partial x_i}\left(\kappa \frac{\partial T}{\partial x_i}\right) d\tau$, T being the local temperature and κ, the thermal conductivity and hence the rate of gain in the thermal energy per unit mass $= \frac{1}{\rho} \frac{\partial}{\partial x_i}\left(\kappa \frac{\partial T}{\partial x_i}\right)$.

Using the first law of thermodynamics for the internal energy per unit mass, δE, gain in the internal energy by heat exchange, δQ per unit mass and the work done on the fluid by amount δW per unit mass, per unit time to give $\delta E = \delta Q + \delta W$, where $\delta Q = \frac{1}{\rho} \frac{\partial}{\partial x_i}\left(\kappa \frac{\partial T}{\partial x_i}\right)$ and $\delta W = \frac{\sigma_{ij}}{\rho} \frac{\partial u_i}{\partial x_j}$, we finally recover

$$\frac{dE}{dt} = \frac{\sigma_{ij}}{\rho} \frac{\partial u_i}{\partial x_j} + \frac{1}{\rho} \frac{\partial}{\partial x_i}\left(\kappa \frac{\partial T}{\partial x_i}\right) \tag{1.46}$$

Substituting the expression for the stress-tensor,

$$\sigma_{ij} = -p\delta_{ij} + \mu_s\left(2e_{ij} - \frac{2}{3}\Delta\delta_{ij}\right) + \eta_b\Delta\delta_{ij} \tag{1.47}$$

we recover the total rate of change of the energy,

$$\frac{dE}{dt} = -\frac{p}{\rho}\Delta + \frac{2\mu_s}{\rho}\left(e_{ij}e_{ij} - \frac{1}{3}\Delta^2\right) + \frac{\eta_b}{\rho}\Delta^2 + \frac{1}{\rho} \frac{\partial}{\partial x_i}\left(\kappa \frac{\partial T}{\partial x_i}\right), \tag{1.48}$$

where the first term on the R.H.S. represents the rate of change of energy due to compression, while the second and third terms stand for dissipation of mechanical energy due to shear and bulk viscosity with last term denoting the rate of gain of heat energy per unit mass. Noting the total energy per unit mass consists of both the internal and kinetic energy of bulk motion, the time rate of change of the total energy (kinetic + internal) can be finally expressed as

$$\frac{d}{dt}\left(u^2 + E\right) = u_i F_i + \frac{1}{\rho} \frac{\partial}{\partial x_j}\left(u_i\sigma_{ij}\right) + \frac{1}{\rho} \frac{\partial}{\partial x_i}\left(\kappa \frac{\partial T}{\partial x_i}\right), \tag{1.49}$$

For a perfect gas we can adopt for the internal energy the expression, $\delta E = C_v \delta T$, C_v being the specific heat at constant volume, to write the expression for time rate of energy as

$$\rho \frac{d}{dT}(C_v T) + p \nabla \cdot \mathbf{u} = \Phi_{vis} + \nabla(\kappa \nabla T), \tag{1.50}$$

where Φ_{vis} is the total viscous dissipation. Using the thermodynamic relation,

$$\delta Q = T \delta S = \delta E + p \delta v \tag{1.51}$$

the temporal change of energy may be expressed as

$$\rho T \frac{dS}{dt} = \Phi_{vis} + \nabla(\kappa \nabla T), \tag{1.52}$$

Expressing the body force per unit mass, $\mathbf{F}$

$$\mathbf{F} = -\nabla \psi, \tag{1.53}$$

where the potential ψ is a function of position independent of time i.e. $\psi = \psi(\mathbf{r})$, to give

$$u_i F_i = -u_i \frac{\partial \psi}{\partial x_i} = -\frac{d\psi}{dt}$$

which represents the rate of change of potential energy for the body force. In the energy balance equation the pressure force appearing in the term

$$-\frac{1}{\rho}\frac{\partial}{\partial x_j}(p u_i \delta_{ij}) = \frac{1}{\rho}\nabla \cdot (p\mathbf{u}) \equiv -\frac{d}{dt}\left(\frac{p}{\rho}\right)$$

Hence, when the potential, ψ and pressure, p are independent of time, we can express the energy equation as,

$$\frac{d}{dt}\left(\frac{1}{2}u^2 + \frac{p}{\rho} + E + \psi\right) = \Phi_{vis} + \nabla(\kappa \nabla T) \tag{1.54}$$

Thus, for an inviscid, non-conducting fluid, we recover the equation

$$\frac{d}{dt}\left(\frac{1}{2}u^2 + \frac{p}{\rho} + E + \psi\right) = 0, \tag{1.55}$$

or $\left(\frac{1}{2}u^2 + \frac{p}{\rho} + E + \psi\right)$ is a constant along the path of the material element-this encapsulates the Bernoulli's equation.

In the absence of viscous and thermal dissipation we recover the thermal equation in the form,

$$\rho C_v \frac{dT}{dt} + p\nabla \cdot \mathbf{u} = 0, \tag{1.56}$$

or for a perfect gas using the relations $p = (R/\mu)\,\rho T = (C_p - C_v)\,\rho T$, $\gamma = C_p/C_v$ and $\frac{1}{p}\frac{dP}{dt} = \frac{1}{\rho}\frac{d\rho}{dt} + \frac{1}{T}\frac{dT}{dt}$ and the mass conservation, $\frac{1}{\rho}\frac{d\rho}{dt} = -\nabla \cdot \mathbf{u}$, we obtain the familiar relation,

$$\frac{1}{p}\frac{dp}{dt} = \frac{\gamma}{\rho}\frac{d\rho}{dt} \tag{1.57}$$

for an adiabatic process; with γ being the ratio of specific heats.

Exercise 1.3 If (ξ, η, ζ) form an orthogonal system of coordinates and if (u_ξ, u_η, u_ζ) represents velocity, then the equation of continuity is

$$\frac{\partial \rho}{\partial t} + \rho(u_\xi s_\xi + u_\eta s_\eta + u_\zeta s_\zeta) + h_1 \frac{\partial}{\partial \xi}(\rho u_\xi) + h_2 \frac{\partial}{\partial \eta}(\rho u_\eta) + h_3 \frac{\partial}{\partial \zeta}(\rho u_\zeta) = 0 \tag{1.58}$$

where

$$1/h_1^2 = (\partial x/\partial \xi)^2 + (\partial y/\partial \eta)^2 + (\partial z/\partial \zeta)^2, \tag{1.59}$$

and similar equations for $1/h_2^2, 1/h_3^2$. The sums of the principal curvatures of the three orthogonal surfaces are denoted by s_ξ, s_η, s_ζ.

Exercise 1.4 Consider a vessel filled with water with an orifice at the bottom. If the vessel is not kept constantly full, motion will not be steady. When the orifice is small compared to the area of the free surface of water the motion may be approximately steady. Find the form of the vessel of revolution with a small orifice at its lowest point so that the surface of the water in it descends uniformly.

Fluid Equations from Kinetic Theory

2

> *Each individual fact, taken by itself, can indeed arouse our curiosity or our astonishment, or be useful to us in its practical applications.*
>
> —Hermann von Helmholtz

The fluid mechanics is one of the most ancient subjects studied by the human civilizations which invariably began along the banks of rivers. The mankind has developed a sound intuitive picture of fluid flows through persistent observations and investigations. Typically, fluids flow are encountered on a wide range of length scales; starting from micro-scales associated with physiological fluid dynamics (blood flows, capillary flows etc) to intermediate scales observed in geophysical fluid dynamics (oceanic flows, river flows etc.) to extremely large scales found in astrophysical flows (solar wind, accretion flow etc.). Although close observation of all such flows reveal complicated local flow patterns, on a large scale one observes a continuous stream of so-called fluid moving from one location to another. On a microscopic level, we know that this stream consists of motion of large number of microscopic particles which may be constantly colliding with each other. Nevertheless, this complicated motion creates an illusion of continuous fluid flow where identity of an individual particle becomes unimportant. We therefore begin our discussion with the validity of continuum approximation which is central for fluid flows.

2.1 Continuum Approximation in Fluid Mechanics

For the purpose of simplicity, we will only consider a classical behaviour of individual particles by neglecting quantum effects. Let us consider a flow of fresh water having a typical mass density of $1 \, \text{g/cm}^3$. Since mass of individual water molecule is around $3 \times 10^{-23} \, \text{g}$, the number of water molecules in a volume of one cubic centimeter is approximately 3×10^{22}. With a similar considerations one can estimate the number density (number of particles per unit volume) of air molecules at the standard temperature and pressure (STP) conditions to be around $10^{19}\,\text{cm}^{-3}$. Thus we find that typical fluid flows encountered in the everyday life have significantly large number densities of microscopic particles [26]. In order to estimate how large should be these number densities for the fluid approximation to be valid, we need to look at a few important length scales associated with the collective motion these particles. These length-scales are described below.

1. **The size of the system (L):** A typical size of the system under consideration forms an important length-scale while defining a fluid flow. This is the macro-scale on which some meaningful measurements of the fluid flow can be made.
2. **The inter-particle distance (d):** Assuming that the particles are uniformly distributed locally, we can estimate the typical spacing between neighboring particles from the local number density, $n(\mathbf{x}, t)$. Here $\mathbf{x}$ and t denote the location and time at which the macroscopic quantities are to be measured. Such an inter-particle distance (d) scales with the number density (n) as

$$d \sim n^{-1/3}. \tag{2.1}$$

Therefore, for the air flow with number density of $10^{19}\,\text{cm}^{-3}$ the inter-particle spacing is of the order of 10^{-7} cm.

3. **The mean free path** (λ_{mfp})**:** Due to significantly large number density, the microscopic particles undergo frequent collisions with each other. A typical distance travelled by these particles between two successive collisions is defined as the mean free path which can be estimated as

$$\lambda_{\text{mfp}} \simeq (n\sigma)^{-1} \tag{2.2}$$

where σ is the collision cross-section. Typically, for hard object collision between two species of radius r_A and r_B, the collision cross-section can be estimated as $\sigma \sim \pi(r_A + r_B)^2$. Therefore, for the case of air flow, taking $\sigma = 10^{-15}\,\text{cm}^2$ and $n = 10^{19}\,\text{cm}^{-3}$, we get $\lambda_{\text{mfp}} \simeq 10^{-4}$ cm.

Thus, we find that the mean free path (λ_{mfp}) is the most dominant *microscopic* length-scale since it is greater than inter-particle distance (d) by orders of magnitude. In order for the continuum or fluid approximation to be valid we need the mean free path between the successive collisions to be much smaller than the typical size of the system i.e.

$$\lambda_{\text{mfp}} \ll L. \tag{2.3}$$

In order to physically understand this situation, let us introduce an intermediate length scale l such that

$$\lambda_{\text{mfp}} \ll l \ll L. \tag{2.4}$$

Since there are large number of collisions over the length scale l, the particles inside the volume element of size $v \sim l^3$ are most glued together due to constant randomization in velocity induced by collisions. We call such volume element as the 'fluid element'. This physically implies that practically all the microscopic particles remain inside the fluid element while they are transported from one location to another under the influence of some external force. Obviously, small fraction of particles present near the surface of the fluid element can still escape out of it due to random collisions. The total velocity $\mathbf{v}$ of an individual particle inside the fluid element is described as

$$\mathbf{v} = \mathbf{u} + \tilde{\mathbf{u}} \tag{2.5}$$

where $\mathbf{u}$ represents a directional mean velocity and $\tilde{\mathbf{u}}$ is the velocity arising out of constant collisions. In such a scenario, it becomes meaningful to compute properties of this fluid element by appropriately averaging over all the particles present inside it by exploiting the randomness of $\tilde{\mathbf{u}}$. Such properties are termed as the 'fluid properties'. In the next section, we derive the governing equations for such properties.

2.2   Derivation of Conservation Laws of Fluid Mechanics

In order to derive the governing equations of a fluid, we begin by introducing a single-particle distribution function $f(\mathbf{x}, \mathbf{v}, t)$ such that

$$f(\mathbf{x}, \mathbf{v}, t)d^3x d^3v \tag{2.6}$$

represents number of particles within phase-space volume $d^3x d^3v$ around $(\mathbf{x}, \mathbf{v})$ at a given time t. This quantity when normalized by total number particles in the system also represents the probability of finding a particle inside the given phase-space volume around $(\mathbf{x}, \mathbf{v})$. Note that this quantity is constructed over a length scale of fluid element (i.e. l). The fluid quantities such as mass density (ρ), mean velocity($\mathbf{u}$) can be derived by appropriate integration of $f(\mathbf{x}, \mathbf{v}, t)$ over velocity-space. For example,

$$\rho(\mathbf{x}, t) = \int_v m f(\mathbf{x}, \mathbf{v}, t) d^3v, \tag{2.7}$$

$$\mathbf{u} = \frac{\int_v m\mathbf{v} f(\mathbf{x}, \mathbf{v}, t) d^3v}{\int_v m f(\mathbf{x}, \mathbf{v}, t) d^3v} = \frac{\int_v m\mathbf{v} f(\mathbf{x}, \mathbf{v}, t) d^3v}{\rho(\mathbf{x}, t)}. \tag{2.8}$$

Here, m denotes mass of an individual particle. In general, mean of any quantity ζ can be calculated by performing appropriate weighted averaging over distribution function as

$$\langle \zeta(\mathbf{x}, t) \rangle = \frac{\int_v \zeta f(\mathbf{x}, \mathbf{v}, t) d^3 v}{\int_v f(\mathbf{x}, \mathbf{v}, t) d^3 v}. \tag{2.9}$$

Thus, in order to derive the time evolution these fluid quantities we need the evolution equation for the distribution function $f(\mathbf{x}, \mathbf{v}, t)$ which can integrated appropriately over the velocity space. This equation, known as the Boltzmann equation [26], is written in the following form:

$$\frac{\partial f}{\partial t} + \frac{\partial f}{\partial x_i} v_i + \frac{\partial f}{\partial v_i} \frac{g_i}{m} = \left(\frac{\partial f}{\partial t} \right)_c. \tag{2.10}$$

where g_i represents i-th component of a body-force (such as gravity) acting on particles. Note that in the above equation, we have used summation notation where repeated indices imply summation over that particular index. The left hand side of the Boltzmann equation represents change in the number of particles inside the given phase-space volume due to phase-space velocity $(d\mathbf{x}/dt, d\mathbf{v}/dt) \equiv (\mathbf{v}, \mathbf{g}/m)$. On the other hand, the right hand side of the equation quantifies the change in the number of particles inside the phase-space volume arising due to inter-particle collisions. While dealing with this term, we assume that the momenta of particles change instantaneously whereas their positions remain unaltered immediately after the collision. The conservation laws for fluid, such as mass, momentum and energy conservation, are derived by taking the velocity moments of the Boltzmann equation. For taking the αth moment, we first multiply the Boltzmann equation by $\mathbf{v}^\alpha$ and then integrate over the entire velocity-space. We now derive such conservation laws by following this procedure.

2.2.1 Mass Conservation Equation

The mass conservation equation for a fluid is obtained by first multiplying the Boltzmann equation by mass (m) of the individual species of the fluid and then integrating over the velocity-space i.e. by taking the zeroth moment of the equation. This leads to

$$\frac{\partial}{\partial t} \left(\int_v mf d^3 v \right) + \int_v m \frac{\partial f}{\partial x_i} v_i d^3 v + \int_v m \frac{\partial f}{\partial v_i} g_i d^3 v = \int_v m \left(\frac{\partial f}{\partial t} \right)_c d^3 v.$$

Rearranging the terms and using the fact that number of particles are conserved in the collisions we get,

$$\frac{\partial}{\partial t} \left(\int_v mf d^3v \right) + \frac{\partial}{\partial x_i} \int_v mf v_i d^3v + \int_v \frac{\partial}{\partial v_i} (f g_i) d^3v - \int_v f \frac{\partial g_i}{\partial v_i} d^3v = 0.$$

The third term in the above equation vanishes by applying divergence theorem (over velocity space) and using the property that $f(\mathbf{x}, \mathbf{v}, t)$ goes to zero as $\mathbf{v} \to \pm\infty$. Also, the last term does not contribute since the body-force does not depend upon the velocity (for example, gravity). This is also true for Lorentz force ($\mathbf{v} \times \mathbf{B}$) since i-th component of this force doesn't depend upon the i-th component of the velocity. Finally, using the definitions of mass density, mean momentum density given by (2.7) and (2.8) respectively we get

$$\frac{\partial \rho}{\partial t} + \frac{\partial}{\partial x_i} (\rho u_i) = 0. \tag{2.11}$$

In the vector form, above equation can be written as

$$\frac{\partial \rho}{\partial t} + \nabla \cdot (\rho \mathbf{u}) = 0. \tag{2.12}$$

This equation can also be expressed as

$$\frac{D\rho}{Dt} = \left[\frac{\partial}{\partial t} + (\mathbf{u} \cdot \nabla) \right] \rho = -\rho \nabla \cdot \mathbf{u} \tag{2.13}$$

where $D/Dt \equiv \partial/\partial t + \mathbf{u} \cdot \nabla$ represents the total derivative, also known as the material derivative operator. This quantity indicates the total change in a quantity with respect to time including the contribution of its advection by the mean fluid velocity ($\mathbf{u}$).

For an incompressible flow, the density of the fluid does not change with time i.e. $D\rho/Dt = 0$. This leads to constrain on the velocity field in the form

$$\nabla \cdot \mathbf{u} = 0. \tag{2.14}$$

In other words, flow of an incompressible fluid is always divergence-free.

2.2.2 Momentum Conservation Equation

The first velocity moment of the Boltzmann equation results in the fluid momentum conservation equation. Therefore, we begin with

$$\frac{\partial}{\partial t} \left(\int_v m v_j f d^3v \right) + \frac{\partial}{\partial x_i} \int_v m v_i v_j f d^3v + \int_v \frac{\partial f}{\partial v_i} v_j g_i d^3v = \int_v m v_j \left(\frac{\partial f}{\partial t} \right)_c d^3v.$$

The collision term on the right hand side vanishes due to conservation of momentum during the collisions. It should be noted that this is true only for a single component system consisting of particles that are identical and indistinguishable. On the other hand, collision term will make finite contribution in a multi-components system having more than one types of particles. For example, in a system with electrons and ions, electron fluid can lose net mean momentum due to the scattering of electrons from ions. Now, using (2.8) and definition of ensemble average given in (2.9), above equation can be rearranged as

$$\frac{\partial}{\partial t}\left(\rho u_j\right) + \frac{\partial}{\partial x_i}\left(\rho\langle v_i v_j\rangle\right) + \int_v \frac{\partial}{\partial v_i}\left(f g_i v_j\right) d^3v - g_i \int_v f \frac{\partial v_j}{\partial v_i} d^3v = 0.$$

Again, the third term drops out using divergence theorem whereas $\partial v_j/\partial v_i = \delta_{ij}$. Thus, we get

$$\frac{\partial}{\partial t}\left(\rho u_j\right) + \frac{\partial}{\partial x_i}\left(\rho\langle v_i v_j\rangle\right) = \rho g_j. \tag{2.15}$$

Now, decomposing the velocity ($\mathbf{v}$) in the mean ($\mathbf{u}$) and fluctuating ($\tilde{\mathbf{u}}$) components, i.e. $\mathbf{v} = \mathbf{u} + \tilde{\mathbf{u}}$, we observe that

$$\langle v_i v_j\rangle = u_i u_j + \langle \tilde{u}_i \tilde{u}_j\rangle. \tag{2.16}$$

Note that, the linear terms in fluctuating velocities vanish during ensemble averaging i.e. $\langle u_i \tilde{u}_j\rangle = \langle u_j \tilde{u}_i\rangle = 0$. Now substituting (2.16) into (2.15) we get

$$\frac{\partial}{\partial t}\left(\rho u_j\right) + \frac{\partial}{\partial x_i}\left(\rho u_i u_j\right) = -\frac{\partial}{\partial x_i}\left(\rho\langle \tilde{u}_i \tilde{u}_j\rangle\right) + \rho g_j. \tag{2.17}$$

Defining stress tensor (σ_{ij}) as

$$\sigma_{ij} = -\rho\langle \tilde{u}_i \tilde{u}_j\rangle = -\int_v m\tilde{u}_i \tilde{u}_j f d^3v = -p\delta_{ij} + \Pi_{ij} \tag{2.18}$$

we get the fluid momentum equation in the form

$$\frac{\partial}{\partial t}\left(\rho u_j\right) + \frac{\partial}{\partial x_i}\left(\rho u_i u_j\right) = -\frac{\partial p}{\partial x_j} + \frac{\partial \Pi_{ij}}{\partial x_i} + \rho g_j. \tag{2.19}$$

Here, $p = \rho\langle \tilde{u}_i \tilde{u}_i\rangle/3$ denotes the pressure which represents the trace of the stress tensor σ_{ij}. On the other hand, the contributions of off-diagonal terms in the stress tensor are incorporated through viscous stress tensor Π_{ij}. Since trace of σ_{ij} is already included in the pressure term, the viscous stress tensor Π_{ij} must be traceless.

Now, using mass conservation equation to simplify the left hand side of Eq. (2.19) we get the final form of momentum conservation equation as

$$\rho \left[\frac{\partial u_j}{\partial t} + u_i \frac{\partial u_j}{\partial x_i} \right] = \frac{\partial \sigma_{ij}}{\partial x_i} + \rho g_j = -\frac{\partial p}{\partial x_j} + \frac{\partial \Pi_{ij}}{\partial x_i} + \rho g_j. \tag{2.20}$$

In the vector form, this equation can be written as

$$\rho \left[\frac{\partial \mathbf{u}}{\partial t} + (\mathbf{u} \cdot \nabla) \mathbf{u} \right] = \nabla \cdot \underline{\underline{\sigma}} + \rho \mathbf{g} = -\nabla p + \nabla \cdot \underline{\underline{\Pi}} + \rho \mathbf{g}. \tag{2.21}$$

Thus, the total acceleration of a fluid element with density ρ depends upon the force $\mathbf{g}$ acting entire fluid element (hence the name 'body-force') and the stresses ($\underline{\underline{\sigma}}$) that act on the surface of the fluid. These stresses are further decomposed into pressure (p) and viscous stresses ($\underline{\underline{\Pi}}$).

2.2.3 Energy Conservation Equation

To obtain the energy conservation equation for a single component fluid, we begin by taking the second velocity moment of the Boltzmann equation i.e. first multiply the equation by $mv_j^2/2$ and then integrate over the velocity space. Therefore,

$$\frac{\partial}{\partial t} \int_v \frac{m}{2} v_j^2 f d^3 v + \frac{\partial}{\partial x_i} \int_v \frac{1}{2} m v_j^2 v_i f d^3 v + \int_v \frac{1}{2} v_j^2 g_i \frac{\partial f}{\partial v_i} d^3 v = \int_v \frac{m}{2} v_j^2 \left(\frac{\partial f}{\partial t} \right)_c d^3 v.$$

Note that here v_j^2 stands for $v_j v_j$ where summation over j is implied. Similar to conservation of mass and momentum equation, the term involving collisions on the right hand side does not contribute for the single component fluid considered here due to conservation of energy during collisions. Now following similar steps used in momentum conservation equation we get

$$\frac{\partial}{\partial t} \left(\frac{\rho}{2} \langle v_j^2 \rangle \right) + \frac{\partial}{\partial x_i} \int_v \frac{1}{2} m v_j^2 v_i f d^3 v = \rho g_k u_k.$$

Decomposing the velocity ($\mathbf{v}$) into mean ($\mathbf{u}$) and fluctuating ($\tilde{\mathbf{u}}$) components we get

$$\frac{\partial}{\partial t} \left[\frac{\rho}{2} \left(u_j^2 + \langle \tilde{u}_j^2 \rangle + 2 \langle u_j \tilde{u}_j \rangle \right) \right] + \frac{\partial}{\partial x_i} \int_v \left[\frac{m}{2} \left(u_j^2 u_i + \tilde{u}_j^2 u_i + 2 u_i u_j \tilde{u}_j \right) \right] f d^3 v$$

$$+ \frac{\partial}{\partial x_i} \int_v \left[\frac{m}{2} \left(u_j^2 \tilde{u}_i + \tilde{u}_j^2 \tilde{u}_i + 2 u_j \tilde{u}_i \tilde{u}_j \right) \right] f d^3 v$$

$$= \rho g_k u_k.$$

Dropping linear terms in fluctuating velocity while averaging and recalling the definition of stress tensor σ_{ij} from (2.18) leads to

$$\frac{\partial}{\partial t}\left[\frac{\rho u_j^2}{2} + \frac{\rho\langle\tilde{u}_j^{\,2}\rangle}{2}\right] + \frac{\partial}{\partial x_i}\left[\frac{\rho u_i u_j^2}{2} + \frac{\rho u_i\langle\tilde{u}_j^{\,2}\rangle}{2} - u_j\sigma_{ij} + F_i\right] = \rho g_k u_k,$$

where F_i denotes the heat flux vector involving third moment of the distribution function. This quantity is defined in the following way:

$$F_i = \int_v \frac{m\tilde{u}_j^{\,2}}{2}\tilde{u}_i f d^3 v \tag{2.22}$$

Now, using mass conservation equation (2.11) the energy equation gets further simplified into

$$\rho\left[\frac{\partial}{\partial t}\left(\frac{u_j^2 + \langle\tilde{u}_j^{\,2}\rangle}{2}\right) + u_i\frac{\partial}{\partial x_i}\left(\frac{u_j^2 + \langle\tilde{u}_j^{\,2}\rangle}{2}\right)\right] + \frac{\partial}{\partial x_i}\left(pu_i - u_j\Pi_{ij}\right) + \frac{\partial F}{\partial x_i} = \rho g_i u_i \tag{2.23}$$

The rate of change of mean kinetic energy can be obtained from momentum conservation equation (2.20) as

$$\rho\left[\frac{\partial}{\partial t}\left(\frac{u_j^2}{2}\right) + u_i\frac{\partial}{\partial x_i}\left(\frac{u_j^2}{2}\right)\right] = -u_i\frac{\partial p}{\partial x_i} + u_j\frac{\partial\Pi_{ij}}{\partial x_i} + \rho g_i u_i \tag{2.24}$$

Using (2.23) and (2.24) we get

$$\rho\left[\frac{\partial}{\partial t}\left(\frac{\langle\tilde{u}_j^{\,2}\rangle}{2}\right) + u_i\frac{\partial}{\partial x_i}\left(\frac{\langle\tilde{u}_j^{\,2}\rangle}{2}\right)\right] = -p\frac{\partial u_i}{\partial x_i} + \Pi_{ij}\frac{\partial u_j}{\partial x_i} - \frac{\partial F_i}{\partial x_i} \tag{2.25}$$

Now, we define specific internal energy (ϵ) and viscous dissipation rate (Ψ) in the following way:

$$\epsilon = \frac{\langle\tilde{u}_j^{\,2}\rangle}{2}, \tag{2.26}$$

$$\Psi = \Pi_{ij}\frac{\partial u_j}{\partial x_i} = \underline{\underline{\boldsymbol{\Pi}}} : \nabla\mathbf{u}. \tag{2.27}$$

Finally, using above definitions the energy conservation equation (2.25) can be written as

$$\rho \left[\frac{\partial \epsilon}{\partial t} + u_i \frac{\partial \epsilon}{\partial x_i} \right] = -p \frac{\partial u_i}{\partial x_i} + \Psi - \frac{\partial F_i}{\partial x_i}. \tag{2.28}$$

In the vector notations, the equation takes the form

$$\rho \left[\frac{\partial \epsilon}{\partial t} + (\mathbf{u} \cdot \nabla) \epsilon \right] = -p \nabla \cdot \mathbf{u} + \Psi - \nabla \cdot \mathbf{F} \tag{2.29}$$

Thus, the first term on the right hand side of the energy conservation equation i.e. $-p \nabla \cdot \mathbf{u}$ represents the change in internal energy due to pdv-work consistent with the first law of thermodynamics. In other words, the average internal energy of fluid particles should increase during compression of the fluid ($\nabla \cdot \mathbf{u} < 0$) whereas it should decrease for the expanding fluid ($\nabla \cdot \mathbf{u} > 0$). It is possible to show that the viscous dissipation rate (Ψ) is a positive definite quantity i.e. it always leads to increase in the average energy. Finally, the last quantity on the right hand side ($\nabla \cdot \mathbf{F}$) represents the change in internal energy due to heat flux through the boundaries of the fluid element. Thus, (2.29) is simply a mathematical statement of the First law of thermodynamics .

Note that we derived (2.29) by considering only translational degrees of freedom through the definition of specific internal energy, $\epsilon = \langle u_j^2 \rangle / 2$. Identifying internal energy per degree of freedom as $k_B T / 2$ where k_B is the Boltzmann constant, for a general molecular with N degrees of freedom the specific internal energy can be written as $\epsilon = N k_B T / 2m$. For example, the internal energy ($m\epsilon$) for a mono-atomic gas is $(3/2) k_B T$ whereas it is $(5/2) k_B T$ for diatomic atomic due to 5 degrees of freedom (3 translational + 2 rotational). Recalling the definition of pressure as $p = \rho \langle \tilde{u}_j^2 \rangle / 3 = \rho k_B T / m$, we can now write $\epsilon = 2p / N\rho$ which can be substituted into (2.29) to get

$$\frac{N}{2} \left[\frac{\partial p}{\partial t} + (\mathbf{u} \cdot \nabla) p \right] = -\left(\frac{N+2}{2} \right) p \nabla \cdot \mathbf{u} + \Psi - \nabla \cdot \mathbf{F}. \tag{2.30}$$

For an isolated system ($\nabla \cdot \mathbf{F} = 0$) without any viscous dissipation ($\Psi = 0$), the above equation takes the form

$$\frac{dp}{p} = \left(\frac{N+2}{N} \right) \frac{d\rho}{\rho} = \gamma \frac{d\rho}{\rho}$$

where $\gamma = (N+2)/N$ is the adiabatic index since it leads to adiabatic equation of state of the form $p/\rho^\gamma = $ constant.

2.3   Closure of the Conservation Laws

The conservation laws, obtained by taking the velocity moments of the Boltzmann equation, are used to describe the change in mass density (ρ), mean momentum (**u**) and pressure (p) of a fluid element as a function of time. For a neutral fluid with N degrees of freedom, these equations are summarized as follows:

$$\frac{D\rho}{Dt} = -\rho\nabla\cdot\mathbf{u},$$

$$\rho\frac{D\mathbf{u}}{Dt} = -\nabla p + \nabla\cdot\underline{\underline{\Pi}} + \rho\mathbf{g}, \qquad (2.31)$$

$$\frac{N}{2}\frac{Dp}{Dt} = -\left(\frac{N+2}{2}\right)p\nabla\cdot\mathbf{u} + \underline{\underline{\Pi}} : \nabla\mathbf{u} - \nabla\cdot\mathbf{F},$$

where the material or convective derivative is defined as $D/Dt \equiv \partial/\partial t + (\mathbf{u}\cdot\nabla) \equiv \partial/\partial t + u_k(\partial/\partial x_k)$. Note that there are five equations in the above set of equations viz. one equation for density and energy each whereas three equations for the momentum vector. On the other hand, the unknowns in this set are $\rho, p, \mathbf{u}, \mathbf{F}, \underline{\underline{\Pi}}$. Clearly the number of unknowns significantly exceed the number of equations available with us !! Therefore, these equations in the present form can't be used to describe the self-consistent evolution of fluid quantities. In order to close the above system of equations, further simplifications are essential which will describe higher rank quantities in terms of low rank fluid quantities present in the system. For example, the rank-2 shear tensor ($\underline{\underline{\Pi}}$) can be expressed in terms of a vector field (rank-1) **u**. Similarly, heat flux vector,**F** (rank-1) can be written in terms of temperature scalar field (rank-0). Such relations are known as the constitutive relations for the fluid.

2.3.1   Constitutive Relations for a Newtonian Fluid

A Newtonian fluid assumes a linear relationship between the viscous stress tensor and local strain rate of the fluid element. This approximation turns out to be quite robust for most of the naturally occurring fluid flows comprising of water, air etc. On the other hand, such a simple linear approximation may not be valid for some complex flows such as viscoelastic flows, paints and polymer flows, blood flows etc. These fluids are recognized as non-Newtonian fluids.

Recalling the definition of stress tensor (σ_{ij}) as

$$\sigma_{ij} = -\rho\langle\tilde{u}_i\tilde{u}_j\rangle = -p\delta_{ij} + \Pi_{ij},$$

we invoke a linear relation between viscous stress tensor (Π_{ij}) and local strain rate which is expressed as rank-2 tensor, $e_{kl} = \partial u_k/\partial x_l$. The local strain rate also characterises the local deformation rate of the fluid element. This second rank tensor can be spilt into symmetric and anti-symmetric part as follows:

$$e_{kl} = \frac{\partial u_k}{\partial x_l} = \frac{1}{2} \left(\frac{\partial u_k}{\partial x_l} + \frac{\partial u_l}{\partial x_k} \right) + \frac{1}{2} \left(\frac{\partial u_k}{\partial x_l} - \frac{\partial u_l}{\partial x_k} \right) \qquad (2.32)$$

The the symmetric part of this tensor corresponds to deformation of the fluid element whereas the anti-symmetric part represents the pure solid body rotation of the fluid element about its principle axes. Since pure rotation of fluid element can not lead to development of stresses in it, only the first part i.e. symmetric part contributes to the stresses. Therefore, we write the general relationship between stress and strain for a Newtonian fluid in the following form:

$$\Pi_{ij} = \frac{\mu_{ijkl}}{2} \left(\frac{\partial u_k}{\partial x_l} + \frac{\partial u_l}{\partial x_k} \right). \qquad (2.33)$$

Here, μ_{ijkl} is the fourth-rank symmetric viscosity tensor since in general two second-rank tensors must be linearly related to each other through a fourth-rank tensor. Note that since viscosity tensor is fourth-rank tensor, there are 81 unknown coefficients associated with it. Therefore prima facie it may seem like we have made the closure problem more difficult by adding significantly more number of unknowns through such a linear approximation. Fortunately, the number of unknowns can be significantly reduced by invoking the isotropic assumption for the fluid transport properties i.e. macroscopic fluid transport properties, such as viscosity and thermal conductivity, are independent of the orientation of the co-ordinate system. This is mainly due to the fact that these properties are the manifestation of some direction independent microscopic random interactions between molecules. Expressing forth rank isotropic tensor μ_{ijkl} in terms of Kronecker-delta tensor (general second rank isotropic tensor) δ_{ij} as

$$\mu_{ijkl} = \lambda \delta_{ij} \delta_{kl} + \mu \left(\delta_{ik} \delta_{jl} + \delta_{il} \delta_{jk} \right) + \gamma \left(\delta_{ik} \delta_{jl} - \delta_{il} \delta_{jk} \right), \qquad (2.34)$$

we can reduce 81 unknowns associated with it to just three scalar quantities λ, μ and γ. This relation is true for any forth rank isotropic tensor. Again noting that the third term in this relation is anti-symmetric, we set $\gamma = 0$ since μ_{ijkl} must also be a symmetric tensor. Therefore, using Eq. (2.33) and (2.34) the stress-strain relationship for a Newtonian fluid can now be expressed as

$$\Pi_{ij} = \left[\frac{\lambda}{2} \delta_{ij} \delta_{kl} + \frac{\mu}{2} \delta_{ik} \delta_{jl} + \frac{\mu}{2} \delta_{il} \delta_{jk} \right] \left(\frac{\partial u_k}{\partial x_l} + \frac{\partial u_l}{\partial x_k} \right).$$

Noting the following relations

$$\frac{\lambda}{2}\delta_{ij}\delta_{kl}\left(\frac{\partial u_k}{\partial x_l} + \frac{\partial u_l}{\partial x_k}\right) = \lambda\delta_{ij}\frac{\partial u_k}{\partial x_k} \qquad (\dots k = l)$$

$$\frac{\mu}{2}\delta_{ik}\delta_{jl}\left(\frac{\partial u_k}{\partial x_l} + \frac{\partial u_l}{\partial x_k}\right) = \frac{\mu}{2}\left(\frac{\partial u_i}{\partial x_j} + \frac{\partial u_j}{\partial x_i}\right) \qquad (\dots i = k, j = l)$$

$$\frac{\mu}{2}\delta_{il}\delta_{jk}\left(\frac{\partial u_k}{\partial x_l} + \frac{\partial u_l}{\partial x_k}\right) = \frac{\mu}{2}\left(\frac{\partial u_j}{\partial x_i} + \frac{\partial u_i}{\partial x_j}\right) \qquad (\dots i = l, j = k)$$

we get following expression for the viscous tensor Π_{ij}

$$\Pi_{ij} = \lambda\delta_{ij}\frac{\partial u_k}{\partial x_k} + \mu\left(\frac{\partial u_i}{\partial x_j} + \frac{\partial u_j}{\partial x_i}\right).$$

Here, μ, λ are called as first and second viscosity coefficients respectively and are typically treated as homogeneous in space.

Having succeeded in reducing the number of unknowns in the momentum equation in a meaningful way, we now turn our attention to the energy equation. The viscous dissipation rate (Ψ) in the energy equation can be written as

$$\Psi = \lambda\frac{\partial u_k}{\partial x_k}\delta_{ij}\frac{\partial u_j}{\partial x_i} + \mu\left(\frac{\partial u_i}{\partial x_j} + \frac{\partial u_j}{\partial x_i}\right)\frac{\partial u_j}{\partial x_i}$$

$$= \lambda\left(\frac{\partial u_k}{\partial x_k}\right)^2 + \frac{\mu}{2}\left(\frac{\partial u_i}{\partial x_j} + \frac{\partial u_j}{\partial x_i}\right)\left[\left(\frac{\partial u_j}{\partial x_i} + \frac{\partial u_i}{\partial x_j}\right) + \left(\frac{\partial u_j}{\partial x_i} - \frac{\partial u_i}{\partial x_j}\right)\right]$$

This gives the expression for the viscous dissipation rate for the Newtonian fluid in the following form:

$$\Psi = \lambda\left(\frac{\partial u_k}{\partial x_k}\right)^2 + \frac{\mu}{2}\left(\frac{\partial u_i}{\partial x_j} + \frac{\partial u_j}{\partial x_i}\right)^2 \qquad (2.35)$$

This quantity is always positive as one would expect from the physical considerations. Finally, we express the rank-3 quantity, heat flux (F_j), in terms of gradient of temperature which is the lower rank quantity using the linear approximation in the following form:

$$F_j = -\kappa\frac{\partial T}{\partial x_j}, \qquad (2.36)$$

where κ is the thermal conductivity of the fluid which assumed to be a scalar (rank zero) for simplicity. Thus, we have succeeded in expressing all the higher rank quantities in terms of lower ranked ones in order to develop self-consistent system of equations to describe the behaviour of a Newtonian fluid.

2.3.2   Equations of State

For solving conservation laws, we need two equations for scalars (ρ, ϵ) and three equations for **u**. The unknown variables in this system of equation are density (ρ), three components of fluid velocity (**u**) along with three thermodynamic quantities viz internal energy (ϵ), pressure (p), temperature (T). Thus, we have total of seven unknowns and five equations to solve. Clearly, in order to solve the system self-consistently we need two additional equations in the form of Equations of State (EOS). With EOS the thermodynamic quantities are expressed in terms of other quantities by a specified equation/relation. For example, the pressure and the specific internal energy can be expressed in terms of density and temperature with $p = \rho RT$ (Ideal Gas equation) and $\epsilon = C_v T$ where R is the ideal gas constant where C_v is the specific heat at constant volume. In general, for complex fluids, where simplified assumptions of kinetic theory does not hold, one needs to get EOS either through experiments or extensive numerical simulations.

2.4   Governing Equations for the Newtonian Fluid

In this section, we summarize the conservation laws for the Newtonian fluid. These equations essentially describe the evolution of fluid properties such as density (ρ), momentum (ρ**u**) and internal energy (ϵ). As we have discussed in the previous section, for the solution of these equations one needs proper prescription of equation of state. In addition, since these are partial differential equations, we need to specify appropriate boundary condition.

2.4.1   Governing Equations for the Compressible Fluid

The final form of conservation equations for mass, momentum and energy for the compressible Newtonian fluid can be expressed as:

$$\frac{\partial \rho}{\partial t} + \frac{\partial}{\partial x_k}(\rho u_k) = 0 \tag{2.37}$$

$$\rho \left[\frac{\partial u_j}{\partial t} + u_k \frac{\partial u_j}{\partial x_k} \right] = -\frac{\partial p}{\partial x_j} + \frac{\partial}{\partial x_j}\left[(\lambda + \mu)\frac{\partial u_k}{\partial x_k} \right] + \mu \frac{\partial^2 u_j}{\partial x_i \partial x_i} \tag{2.38}$$

$$\rho \left[\frac{\partial \epsilon}{\partial t} + u_k \frac{\partial \epsilon}{\partial x_k} \right] = -p\frac{\partial u_k}{\partial x_k} + \frac{\partial}{\partial x_k}\left(\kappa \frac{\partial T}{\partial x_k} \right) + \lambda \left(\frac{\partial u_k}{\partial x_k} \right)^2 + \frac{\mu}{2}\left(\frac{\partial u_i}{\partial x_j} + \frac{\partial u_j}{\partial x_i} \right)^2 \tag{2.39}$$

In vector form, these equations are written as follows:

$$\frac{\partial \rho}{\partial t} + \nabla \cdot (\rho \mathbf{u}) = 0 \tag{2.40}$$

$$\rho \left[\frac{\partial \mathbf{u}}{\partial t} + (\mathbf{u} \cdot \nabla) \mathbf{u} \right] = -\nabla p + \rho \mathbf{g} + \mu \nabla^2 \mathbf{u} + \nabla \left[(\lambda + \mu) \nabla \cdot \mathbf{u} \right]. \tag{2.41}$$

$$\rho \left[\frac{\partial \epsilon}{\partial t} + (\mathbf{u} \cdot \nabla) \epsilon \right] = -p \nabla \cdot \mathbf{u} + \nabla (\kappa \nabla T) + \Psi \tag{2.42}$$

where Ψ is the viscous dissipation rate, defined in Eq. (2.35). Above equations are supported through EOS with

$$p = p(\rho, T) \tag{2.43}$$

$$\epsilon = \epsilon(\rho, T) \tag{2.44}$$

2.4.2   Governing Equations for an Incompressible Fluid

For an incompressible fluid, we assume that the density is perfectly constant i.e. $d\rho/Dt = 0$. The behaviour of an incompressible fluid, in the vector form, is described through following equations.

$$\nabla \cdot \mathbf{u} = 0 \tag{2.45}$$

$$\rho \left[\frac{\partial \mathbf{u}}{\partial t} + (\mathbf{u} \cdot \nabla) \mathbf{u} \right] = -\nabla p + \rho \mathbf{g} + \mu \nabla^2 \mathbf{u}. \tag{2.46}$$

$$\rho \left[\frac{\partial \epsilon}{\partial t} + (\mathbf{u} \cdot \nabla) \epsilon \right] = \nabla (\kappa \nabla T) + \Psi \tag{2.47}$$

Note that the term representing work done by pressure is absent in the energy equation due to the divergence free nature of the fluid velocity field. The energy equation gets decoupled with the momentum equation when the terms representing viscous dissipation and heat flux are negligible. Although these assumptions may be valid for common fluids; for highly viscous fluids viscous heating can play a significant role in overall dynamics. In general, we observe that the first two equations (mass and momentum conservation) in the above system of equations form a closed set involving four equations and four unknowns ($\mathbf{u}$, p). In this sense, the energy equation can be termed as redundant. In order to incorporate divergence-free velocity constraint, we take the divergence of the momentum equation to get

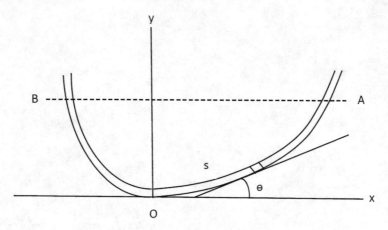

Fig. 2.1 Water in a bent tube is at an equilibrium level, marked by AB. A small perturbation sets water into small oscillations, whose time-period is (2.49)

the Poisson equation for the pressure in the form

$$\nabla^2 p = \nabla \cdot [\mathbf{g} - (\mathbf{u} \cdot \nabla)\,\mathbf{u}].$$ (2.48)

Therefore, in the incompressible fluid, the pressure adjusts itself to the bodyforce (**g**) and the velocity field (**u**) satisfy the divergence free velocity constraint.

Exercise 2.1 Let us consider a flow field with components of velocity at a point (x, y, z), $(-y/r^2, x/r^2, 0)$ where r denotes the distance from the z-axis. Verify that the equation of continuity is satisfied. The fact that equation of continuity is satisfied assures that this represent a possible motion. Find the streamlines of the flow.

Exercise 2.2 Let us consider a water in a bent uniform tube in a vertical plane (Fig. 2.1). Let h be the height of the equilibrium level, AB, above O. The inclinations of the tube at A and B with respect to horizontal are α and β respectively. The inclination is assumed to be θ at a distance s from O. Denote OA by 'a' and OB by 'b'. At a certain time t, water is displaced a small distance x along the tube about the equilibrium position. With equation of continuity, $\partial u/\partial s = 0$, and equation of motion, $\partial u/\partial t = -g \sin \theta - (1/\rho)\partial p/\partial s$, the oscillation period is [96]

$$T = 2\pi \sqrt{\frac{a+b}{g(\sin \alpha + \sin \beta)}}$$ (2.49)

Exercise 2.3 Why two pulsating spheres in a liquid attract each other if they are always in same phase [103].

Vorticity

> *Big whirls have little whirls, That feed on their velocity; And little whirls have lesser whirls, And so on to viscosity.*
>
> —*Lewis Fry Richardson*

In this chapter we will discuss an important fluid property called vorticity. Mathematically, vorticity is defined as the curl of velocity of the fluid i.e.

$$\boldsymbol{\omega} = \nabla \times \mathbf{u} \tag{3.1}$$

Physically, vorticity is related to the circulation of fluid along a closed contour inside the flow. It is also related to rotation of the fluid and can be numerically calculated as the twice the angular speed of rotation of the fluid element about its own axes. The circulation around any closed material curve embedded in the fluid is given by

$$C = \oint \boldsymbol{u}.d\boldsymbol{\ell}. \tag{3.2}$$

Applying Stokes' theorem, we get the relation between circulation and vorticity as

$$C = \int_A \boldsymbol{\omega}.\mathbf{n}\,dA, \tag{3.3}$$

where $\mathbf{n}$ is the unit vector normal to the surface over which integration is carried out. Note that the vorticity is a vector and at every point in the flow-field represented by $\boldsymbol{u}(\mathbf{x}, t)$ there is an associated vorticity. Thus, just the way streamline is defined as the line whose tangent at the point gives the direction of local fluid velocity, we can define a vortex line whose tangent will be in the direction of local vorticity vector.

© The Author(s), under exclusive license to Springer Nature Switzerland AG 2022
S. R. Jain et al., *A Primer on Fluid Mechanics with Applications*,
https://doi.org/10.1007/978-3-031-20487-6_3

Fig. 3.1 Vortex tube inside a
fluid flow: The red arrows
represent vorticity vector
whereas the black arrows are
in the direction of normal to
the surface

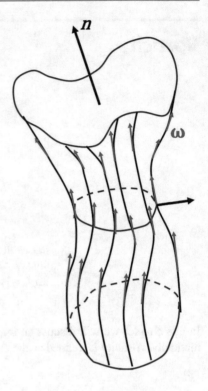

Similarly, we can visualize a vortex tube to be formed of vortex lines emanating
from a close curve as shown in Fig. 3.1. From the definition given in Eq. (3.1), we
see that the vorticity is a divergence-free vector i.e.

$$\nabla \cdot \boldsymbol{\omega} = 0. \tag{3.4}$$

This condition immediately reminds us about the fundamental constrain on a
magnetic field vector ($\mathbf{B}$) in the electrodynamics i.e. $\nabla \cdot \mathbf{B} = 0$. Therefore, analogous
to magnetic field lines, the vortex lines form either closed loops or terminate on solid
or free surface boundaries. Also, applying this constrain on a volume v enclosed by
a surface S we get a relation similar to that of magnetic flux conservation in the
electrodynamics.

$$\int_V \nabla \cdot \boldsymbol{\omega} dV = \int_S \boldsymbol{\omega} \cdot \mathbf{n} dA = 0 \tag{3.5}$$

This analogy with magnetic field is extremely useful in analysing vorticity dynamics
and will be discussed in detail later.

3.1 Vorticity Equation

We begin with the momentum equation for a Newtonian fluid, derived in previous chapters.

$$\left[\frac{\partial \mathbf{u}}{\partial t} + (\mathbf{u} \cdot \nabla)\,\mathbf{u}\right] = -\frac{1}{\rho}\nabla p + \mathbf{g} + \nu\nabla^2\mathbf{u} + \frac{1}{\rho}\nabla\left[(\lambda + \mu)\,\nabla \cdot \mathbf{u}\right]$$

where ν is the kinematic viscosity defined as $\nu = \mu/\rho$. Taking curl of this equation, for a conservative force we get

$$\frac{\partial}{\partial t}\,(\nabla \times \mathbf{u}) + \nabla \times [\mathbf{u} \cdot \nabla\mathbf{u}] = -\nabla\times\left(\frac{\nabla p}{\rho}\right) + \nu\nabla^2\,(\nabla \times \mathbf{u}) + \nabla\times\left(\frac{(\lambda + \mu)}{\rho}\nabla\,(\nabla \cdot \mathbf{u})\right)$$

Noting $(\mathbf{u} \cdot \nabla)\mathbf{u} = \nabla(\mathbf{u}^2/2) - \mathbf{u} \times (\nabla \times \mathbf{u})$ and the definition of vorticity, $\boldsymbol{\omega} = \nabla \times \mathbf{u}$ we get

$$\frac{\partial \boldsymbol{\omega}}{\partial t} = \nabla \times (\mathbf{u} \times \boldsymbol{\omega}) + \frac{\nabla\rho \times \nabla p}{\rho^2} + \tilde{\nu}\frac{\nabla\rho \times \nabla\,(\nabla \cdot \mathbf{u})}{\rho^2} + \nu\nabla^2\boldsymbol{\omega},$$

where $\tilde{\nu} = (\lambda+\mu)/\rho$ is the corresponding kinematic viscosity. Using $\nabla\times(\boldsymbol{u} \times \boldsymbol{\omega}) = (\boldsymbol{\omega}.\nabla)\boldsymbol{u} - (\boldsymbol{u}.\nabla\boldsymbol{\omega}) + \boldsymbol{u}\nabla.\boldsymbol{\omega} - \boldsymbol{\omega}\nabla.\boldsymbol{u}$ and noting $\nabla \cdot \boldsymbol{\omega} = 0$ we can write the above equation as

$$\frac{D\boldsymbol{\omega}}{Dt} = (\boldsymbol{\omega} \cdot \nabla)\boldsymbol{u} - \boldsymbol{\omega}\nabla \cdot \boldsymbol{u} + \frac{\nabla\rho \times \nabla p}{\rho^2} + \tilde{\nu}\frac{\nabla\rho \times \nabla\,(\nabla \cdot \boldsymbol{u})}{\rho^2} + \nu\nabla^2\boldsymbol{\omega}, \qquad (3.6)$$

The generation of vorticity due to non-uniform fluid density can be seen from third and fourth terms of the right hand side of the above equation. These terms are effective only when the gradients in pressure/velocity field divergence are not aligned with the gradients in the density. In order to understand the mechanism of vorticity generation, we consider a situation when density and pressure gradients are exactly perpendicular to each other as shown in Fig. 3.2. In this situation, each fluid element experiences a torque due to horizontal pressure force acting on a vertically varying mass. This torque is produced by the pressure force passing through the geometric centre for the fluid element which is misaligned with its centre of mass due to the non-uniform density. The collective response of the fluid in such situation is through generation of large-scale flows as depicted in the figure. This is one of the dominant mechanism of large-scale flows in the geophysical fluid dynamics where lateral temperature gradients interact with the vertical density gradients arising due to the stratification by gravity. Similar situation arises when density gradient is not aligned with the gradient of divergence of the velocity field. This mechanism is effective when the shock wave impinges on a density stratified fluid environment. On the other hand, in case of barotropic flow where pressure is only a function of density, the $\nabla\rho \times \nabla p$ term is absent. Similarly, for an incompressible fluid,

Fig. 3.2 Generation of
vorticity with non-parallel
density and pressure
gradients.Each fluid element
experiences the tilt due to
differential torque which in
turn leads to large-scale
circulation

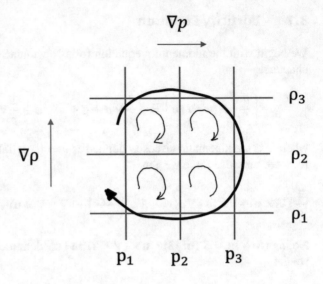

the second term involving $\nabla(\nabla \cdot \mathbf{u})$ is absent. Note that these are the only terms
in the vorticity equation which are independent of ω. Therefore, these terms can
potentially generate vorticity even if it is absent at initial time. Hence, these terms
are considered as the source terms in the vorticity equation.

In the absence of source terms, the equation reduces to

$$\frac{D\omega}{Dt} = (\omega \cdot \nabla)\mathbf{u} - \omega\nabla \cdot \mathbf{u} + \nu\nabla^2\omega, \tag{3.7}$$

The last term in this equation represents the diffusion of vorticity due to viscous
torque arising due to the varying viscous stresses acting the surface of the fluid
element. The diffusion of vorticity due to viscosity is analogous to the viscous
diffusion of momentum inside the fluid. Neglecting this term and using $D\rho/Dt = -\rho(\nabla \cdot \mathbf{u})$, we get the vorticity equation for inviscid compressible fluid as

$$\frac{D}{Dt}\left(\frac{\omega}{\rho}\right) = \left(\frac{\omega}{\rho} \cdot \nabla\right)\mathbf{u}. \tag{3.8}$$

This equation shows the transport of vorticity per unit mass by the fluid velocity.

3.1.1   Vorticity Equation for the Incompressible Fluid

For an incompressible fluid, we get the vorticity equation in the standard form as

$$\frac{\partial\omega}{\partial t} = \nabla \times (\mathbf{u} \times \omega) + \nu\nabla^2\omega, \tag{3.9}$$

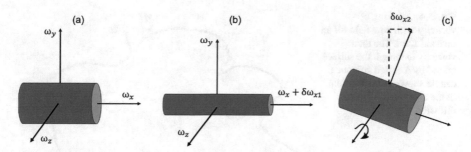

Fig. 3.3 Interpretation of $(\boldsymbol{\omega}\cdot)$ $\mathbf{u}$ term in terms of stretching and tilting of the fluid element. When the original fluid element (**a**) stretches along x-direction as shown in (**b**), its momenta of inertia changes causing change in vorticity along x-direction. For example, in the case of cylindrical fluid element, stretching of causes reduction in the radius and hence increased rotation speed along x-direction. Similarly tilting of fluid element due to rotation about y and z axes produces change in vorticity in x-direction as shown in (**c**)

Expanding the curl of a vector product and following similar steps as the previous section, we can re-write this as

$$\frac{D\boldsymbol{\omega}}{Dt} = (\boldsymbol{\omega}.\nabla)\boldsymbol{u} + \nu\nabla^2\boldsymbol{\omega}. \tag{3.10}$$

This equation shows that total change in the vorticity in an incompressible fluid happens either through viscous diffusion or through stretching and tilting of the fluid element. The interpretation of the term $(\boldsymbol{\omega}.\nabla)\boldsymbol{u}$ in terms of stretching and tilting of the fluid element can be seen with help of Fig. 3.3. Consider a cylindrical fluid element (a) getting stretched along x-direction through a velocity shear $\partial u/\partial x$ as shown in Fig. (b). Due to this stretching, the moment of inertia of the fluid element changes leading to change in the x-component of the vorticity through conservation of angular momentum. This change can be written as $\delta\omega_{x1} = \omega_x \partial u/\partial x$. On the other hand, this component of vorticity also changes when the fluid element tilts due rigid body rotation about y and z axes as shown it Fig. (c). It can also be shown that this contribution can be written as $\delta\omega_{x2}+\delta\omega_{x3} = \omega_y \partial v/\partial y+\omega_z \partial w/\partial z$. This overall contribution of stretching and tilting of fluid element to change in x-component of vorticity can be written as $(\boldsymbol{\omega} \cdot \nabla)\,u = \omega_x \partial u/\partial x + \omega_y \partial v/\partial y + \omega_z \partial w/\partial z$. The vortex stretching term plays an important role in the description of a turbulent fluid. For example, the absence of this term in two dimensional flows leads to qualitative different behaviour of two dimensional turbulent flows compared to that of three dimensional flows.

A typical timescale for vorticity diffusion over length L can be estimated as ν/L^2 whereas the advection of vorticity due to fluid velocity U happens over the time-scale of L/U. When the diffusive timescale is dominant, i.e. significantly shorter than the advection timescale, one can neglect the terms scaling with U/L in the

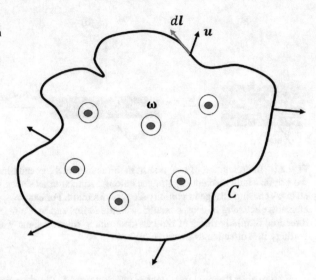

Fig. 3.4 Freezing of vorticity with flow field for an inviscid fluid. The total vorticity ω though the surface bound by a closed contour C can be shown to be invariant in the absence of viscous diffusion

vorticity equation leading to the standard diffusion equation in the form

$$\frac{\partial \omega}{\partial t} = \nu \nabla^2 \omega. \tag{3.11}$$

On the other hand, in the advection dominant regime the vorticity equation reduces to

$$\frac{D\omega}{Dt} = (\omega.\nabla)u. \tag{3.12}$$

This can be interpreted as if the vorticity is frozen to the fluid flow. This can be seen by looking at total change in the vorticity over a closed contour as shown in Fig. 3.4. Therefore, total rate of change of vorticity through the surface bound by a closed contour C is given by

$$\frac{D}{Dt} \oint (\omega \cdot d\mathbf{A}) = \oint (\omega \cdot \nabla)\, u. \cdot d\mathbf{A}$$

Expressing differential area element ($d\mathbf{A}$) in terms of differential length element ($d\mathbf{l}$) as $d\mathbf{A} = (\mathbf{u} \times d\mathbf{l})\, dt$, we get

$$\frac{D}{Dt} \oint (\omega \cdot d\mathbf{A}) = \oint (\omega \cdot \nabla)\, u. \cdot [\mathbf{u} \times d\mathbf{l}]\, dt = \oint d\mathbf{l} \cdot [(\omega \cdot \nabla)\, u \times u]\, dt = 0 \tag{3.13}$$

Thus, total vorticity through the surface remains invariant with the flow field i.e. even if the contour C gets distorted by the flow field $\mathbf{u}$, the flux of vorticity $\oint (\boldsymbol{\omega} \cdot d\mathbf{A})$ remains same through the surface bound by the distorted contour as if it is frozen with the velocity field.

Exercise 3.1 Beginning with Maxwell's equations for electromagnetism, show that there is an equation satisfied by the evolution of magnetic field which is analogous to (3.9):

$$\frac{\partial \mathbf{B}}{dt} = \nabla \times (\mathbf{u} \times \mathbf{B}) + \eta \nabla^2 \mathbf{B} \tag{3.14}$$

where $\eta = c^2/(4\pi \mu \sigma)$.

Analogously, for the magnetic case, if we put $\eta = 0$, we have

$$\frac{d}{dt} \frac{\mathbf{B}}{\rho} = \left(\frac{\mathbf{B}}{\rho}.\nabla\right) \mathbf{u}. \tag{3.15}$$

If there is no change of $\mathbf{u}$ in the direction of $\boldsymbol{\omega}$ $(\mathbf{B})$, then $(\boldsymbol{\omega}/\rho)$ $((\mathbf{B}/\rho))$ remains constant in time.

3.2 Kelvin's Circulation Theorem

The rate of change of circulation is

$$\frac{dC}{dt} = \oint \frac{d\mathbf{u}}{dt}.d\boldsymbol{\ell} + \oint \mathbf{u}.(d\boldsymbol{\ell}.\nabla)\mathbf{u}$$

$$= \oint \frac{d\mathbf{u}}{dt}.d\boldsymbol{\ell} + \oint d\boldsymbol{\ell}.(\mathbf{u}.\nabla)\mathbf{u}$$

$$= \oint \mathbf{F}.d\boldsymbol{\ell} + \oint d\boldsymbol{\ell}.\nabla \left(-\frac{p}{\rho} + \frac{u^2}{2}\right) + v \oint \nabla^2 \mathbf{u}.d\boldsymbol{\ell}. \tag{3.16}$$

$\mathbf{F}$ can be written as $-\nabla\psi$, thus in

$$\frac{dC}{dt} = \oint d\boldsymbol{\ell}.\nabla \left(-\psi - \frac{p}{\rho} + \frac{u^2}{2}\right) + v \oint \nabla^2 \mathbf{u}.d\boldsymbol{\ell}, \tag{3.17}$$

the first term on RHS is zero because ψ, p/ρ, u^2 are single-valued functions. Thus,

$$\frac{dC}{dt} = \nu \oint \nabla^2 \boldsymbol{u}.d\boldsymbol{\ell}$$

$$= -\nu \oint d\boldsymbol{\ell}.\nabla \times \boldsymbol{\omega}$$

$$= -\nu \int \epsilon_{ijk} \frac{\partial \omega_k}{\partial x_j} d\ell_i. \qquad (3.18)$$

The time rate of change of circulation is thus related to the flux of vorticity as

$$\frac{d}{dt} \oint \boldsymbol{u}.d\boldsymbol{\ell} = \int_S (\nabla \times \boldsymbol{u}).\hat{n}dS$$

$$= \int_S \boldsymbol{\omega}.\hat{n}dS = -\nu \oint (\nabla \times \boldsymbol{\omega}).d\boldsymbol{\ell}. \qquad (3.19)$$

Thus, the flux of vorticity across an open surface bounded by a closed wire, Γ cannot be generated in any way and can solely change solely by the process of viscous diffusion.

Kelvin's circulation theorem states that circulation round any closed curve is invariant for an inviscid ($\nu = 0$) in an incompressible fluid. The theorem for an incompressible, inviscid fluid states that if the circulation around a closed curve Γ is initially zero, then the irrotational motion (i.e., vorticity free) is preserved for all times.

3.3   Charged Fluids and Magnetovorticity

Maxwell's equations have to be supplemented with generalized Ohm's law. Consider a completely ionized has (Hydrogen) with ion number density n_i and electron density n_e.

We define magnetovorticity

$$\boldsymbol{\omega}_m = \boldsymbol{\omega} + \frac{e\boldsymbol{B}}{m_i c}. \qquad (3.20)$$

The time-evolution of $\boldsymbol{\omega}_m$ is

$$\frac{\partial \boldsymbol{\omega}_m}{\partial t} = \nabla \times (\boldsymbol{u} \times \boldsymbol{\omega}_m). \qquad (3.21)$$

We want to derive the induction equation by complementing Maxwell's equations (ignoring the displacement current) with generalized Ohm's law,

$$\mathbf{j} = \sigma \left(\mathbf{E} + \frac{\mathbf{v} \times \mathbf{B}}{c} - \frac{\mathbf{j} \times \mathbf{B}}{c n_e e} \right). \tag{3.22}$$

Curl of the current density, upon using the Maxwell equation for curl of electric field gives

$$\nabla \times \mathbf{j} = \sigma \left[-\frac{1}{c}\frac{\partial \mathbf{B}}{\partial t} + \frac{1}{c}\nabla \times (\mathbf{v} \times \mathbf{B}) - \frac{1}{c n_e e}\nabla \times (\mathbf{j} \times \mathbf{B}) \right]. \tag{3.23}$$

Using $\mathbf{j} = (c/4\pi \mu)\nabla \times \mathbf{B}$, and rearranging the terms, we have

$$\frac{\partial \mathbf{B}}{\partial t} = \nabla \times (\mathbf{v} \times \mathbf{B}) - \frac{c}{4\pi n_i e \mu}\nabla \times ((\nabla \times \mathbf{B}) \times \mathbf{B}) + \eta \nabla^2 \mathbf{B}. \tag{3.24}$$

The equation of motion is

$$\frac{\partial \mathbf{u}}{\partial t} + (\mathbf{u}.\nabla)\mathbf{u} = -\nabla \left(\frac{p}{\rho} \right) + \frac{\mathbf{j} \times \mathbf{B}}{\rho c} + \nu^2 \nabla^2 \mathbf{u}. \tag{3.25}$$

Curl of this equation entails an equation for vorticity:

$$\frac{\partial \boldsymbol{\omega}}{\partial t} = -\nabla \times (\mathbf{u}.\nabla)\mathbf{u} - \nabla \times (\nabla(p/\rho)) + \nabla \times (\mathbf{j} \times \mathbf{B})/c + \nu^2 \nabla(\nabla.\boldsymbol{\omega}) \tag{3.26}$$

Using (3.24), (3.26), we obtain in the absence of diffusive terms:

$$\frac{\partial}{\partial t}\left(\boldsymbol{\omega} + \frac{e\mathbf{B}}{m_i c} \right) = \nabla \times \left[\mathbf{u} \times \left(\boldsymbol{\omega} + \frac{e\mathbf{B}}{m_i c} \right) \right], \tag{3.27}$$

an equation for magnetovorticity, $\boldsymbol{\omega}_m$, which is an invariant entity in the absence of diffusion.

Exercise 3.2 Consider a steady-state inviscid fluid experiencing no body-force. For such a fluid, show that isentropic flows are always irrotational and vice versa provided that the stagnation enthalpy (h_0) is uniform throughout the fluid. The stagnation enthalpy is defined as $h_0 = h + u^2/2$ where h is the specific enthalpy and u is the magnitude of fluid velocity.

Exercise 3.3 Consider a sink vortex formed due to drain located at the origin. Assuming the fluid to be incompressible and inviscid, explain the hyperbolic depression (in height) in the central part of the sink.

Potential Flows in Two Dimensions

4

> *If people do not believe that mathematics is simple, it is only because they do not realize how complicated life is.*
>
> —*John von Neumann*

The Navier-Stokes equations are greatly simplified when the fluid is both incompressible and irrotational. Such fluids are typically referred as the 'ideal fluids'. For this special case, the entire velocity field can be expressed in terms of scalar functions. In this chapter, we will deal with such a description which is extremely powerful for developing physical intuition of fluid flows.

We begin by considering a two-dimensional incompressible and inviscid (viscosity, $\mu = 0$) fluid. For an incompressible fluid, we have seen that the velocity field ($\mathbf{u}$) should be divergence free i.e. $\nabla \cdot \mathbf{u} = 0$. For a two-dimensional flow, in Cartesian coordinates this condition is expressed as

$$\frac{\partial u}{\partial x} + \frac{\partial v}{\partial y} = 0 \tag{4.1}$$

Using the fact that $\nabla \cdot (\nabla \times \mathbf{A}) = 0$ for any vector $\mathbf{A}$, we can express the velocity field $\mathbf{u}$ as

$$\mathbf{u} = \nabla \times \left[\psi(x, y)\hat{z} \right], \tag{4.2}$$

where $\psi(x, y)$ is called as the stream function. The components of velocity can now be written as

$$u = \frac{\partial \psi}{\partial y}, \quad v = -\frac{\partial \psi}{\partial x}. \tag{4.3}$$

© The Author(s), under exclusive license to Springer Nature Switzerland AG 2022
S. R. Jain et al., *A Primer on Fluid Mechanics with Applications*,
https://doi.org/10.1007/978-3-031-20487-6_4

Note that along a curve $\psi(x, y) = C$ where C is a constant, we have

$$d\psi = \frac{\partial \psi}{\partial x}dx + \frac{\partial \psi}{\partial y}dy = -vdx + udy = 0,$$

where we have used the definition of stream function from (4.3). Therefore, the slope of the streamline is given by

$$\left(\frac{dy}{dx}\right)_\psi = \frac{v}{u}. \tag{4.4}$$

Thus, contours of constant streamfunction are in the direction of streamlines and give us the information about the flow direction at any given point (x, y).

Now we consider the vorticity (ω) evolution equation, derived in the previous chapter, for an incompressible, inviscid $(\mu = 0)$ fluid in the following form:

$$\frac{D\omega}{Dt} = (\omega \cdot \nabla) \mathbf{u}$$

This equation shows that an incompressible, inviscid fluid remains irrotational provided it has zero initial vorticity. Such a fluid must satisfy following constrain on the its velocity field:

$$\omega = \nabla \times \mathbf{u} = 0 \tag{4.5}$$

Therefore, the velocity field $\mathbf{u}$ must be expressible in terms of a scalar potential function ϕ as

$$\mathbf{u} = -\nabla \phi. \tag{4.6}$$

The individual velocity components in two-dimensional Cartesian system can now be written as

$$u = \frac{\partial \phi}{\partial x}, \quad v = \frac{\partial \phi}{\partial y} \tag{4.7}$$

The equation for constant potential line can be derived by setting

$$d\phi = \frac{\partial \phi}{\partial x}dx + \frac{\partial \phi}{\partial y}dy = -(udx + vdy) = 0. \tag{4.8}$$

This leads to the slope of the constant potential line as

$$\left(\frac{dy}{dx}\right)_\psi = -\frac{u}{v}. \tag{4.9}$$

Comparing Eqs. (4.4) and (4.9), we conclude that the streamlines and constant potential lines are perpendicular to each other.

For an ideal fluid, combining incompressibility and irrotationality conditions lead to

$$\nabla^2 \psi = \nabla^2 \phi = 0. \tag{4.10}$$

Thus, both stream-function and potential function satisfy the Laplace equation for such a fluid. Therefore, each term in the general solution of Laplace equation can in-principle represent a certain type of flow pattern of an ideal fluid. In addition, exploiting the linearity of this equation, we can construct complex flow patterns by combining two or more solutions using the principle of superposition.

The streamfunction and potential function in the polar co-ordinate system (R, θ) can be obtained in the similar way. For stream function, we consider a continuity equation

$$\nabla \cdot \boldsymbol{u} = \frac{1}{r} \frac{\partial}{\partial r} \left(r \frac{\partial u_r}{\partial r} \right) + \frac{1}{r} \frac{\partial u_\theta}{\partial \theta} = 0. \tag{4.11}$$

Therefore, the velocity components (u_r, u_θ) are expressed in terms of streamfunction as

$$u_r = \frac{1}{r} \frac{\partial \psi}{\partial \theta}, \quad u_\theta = -\frac{\partial \psi}{\partial r}. \tag{4.12}$$

Similar, for two-dimensional polar flows the vorticity vector points in the z-direction. In addition, due to irrotationality we get

$$(\nabla \times \boldsymbol{u})_z = \frac{1}{r} \frac{\partial}{\partial r} \left(r \frac{\partial u_\theta}{\partial r} \right) - \frac{1}{r} \frac{\partial u_r}{\partial \theta} = 0. \tag{4.13}$$

This leads to expression of velocity components in terms of potential function as

$$u_r = \frac{\partial \phi}{\partial r}, \quad u_\theta = -\frac{1}{r} \frac{\partial \phi}{\partial \theta}. \tag{4.14}$$

Again, one can readily verify that both potential and streamfunction satisfy the Laplace equation in the polar co-ordinate system.

Exercise 4.1 For an ideal fluid show that the potential function ϕ satisfied following equation

$$\frac{\partial \phi}{\partial t} + \frac{|\nabla \phi|^2}{2} + \frac{p}{\rho} = \text{constant} \tag{4.15}$$

where p, ρ are the pressure and density respectively. Neglect the body force on the fluid.

4.1   Use of Complex Analysis in Potential Flow Theory

For two-dimensional potential flows, the streamfunction and potential function are related through Cauchy-Riemann equations

$$u = \frac{\partial \phi}{\partial x} = \frac{\partial \psi}{\partial y}$$

$$v = \frac{\partial \phi}{\partial y} = -\frac{\partial \psi}{\partial x}$$

Therefore, we introduce a complex potential function in terms of an analytic function $w(z)$ as

$$W(z) = \phi(x, y) + i\psi(x, y) \tag{4.16}$$

where $z = x + iy$. Since $W(z)$ is an analytic function, the real (ϕ) and imaginary (ψ) parts of this function will automatically satisfy the Laplace equation. Therefore, every complex analytic function can be used to represent a flow pattern of an ideal fluid. Also, just as we defined the real velocity as the gradient of real potential, we express complex velocity as

$$\frac{dW}{dz} = u - iv. \tag{4.17}$$

Note that the magnitude of fluid velocity can be obtained from

$$\left(\frac{dW}{dz}\right)\left(\frac{dW}{dz}\right)^* = u^2 + v^2, \tag{4.18}$$

where $(dW/dz)^*$ represents the complex-conjugate of the fluid velocity.

The velocity components in the Cartesian coordinates system (u, v) are written in terms of polar coordinates system as

$$u = u_r \cos\theta - u_\theta \sin\theta,$$

$$v = u_r \sin\theta + u_\theta \cos\theta \tag{4.19}$$

4.1.1   Complex Potentials of the Form $W(z) = CZ^n$

By proper choice of analytic function $W(z)$ for the complex potential, various flow patterns of two-dimensional ideal flow can be constructed. We begin by considering simplest analytic function of type

$$W(z) = Cz^n, \quad n \geq 0, \tag{4.20}$$

for an integer number n. Clearly $n = 0$ is not of interest, since it represents $W(z) = C$, a constant whose differentiation will give us zero flow velocity. Note that clever choice of a constant complex potential can be used to create an equipotential surface to satisfy specific boundary conditions without disturbing the overall flow pattern, as will be demonstrated later.

4.1.1.1 Construction of Uniform Flows
The choice of $n = 1$ leads to the complex potential of the form

$$W(z) = Cz. \tag{4.21}$$

The corresponding complex velocity is given by

$$\frac{dW}{dz} = C = U_0 - iV_0 = \sqrt{U_0^2 + V_0^2} e^{-i\theta_v}, \tag{4.22}$$

where θ_v represents angle of inclination of the uniform flow with the horizontal and is given by

$$\tan \theta_v = \frac{V_0}{U_0}. \tag{4.23}$$

From Eqs. (4.17) and (4.22), we see that when C is a real number we have a uniform horizontal flow with speed U_0 whereas for purely imaginary C will lead to a vertical flow of magnitude V_0. In general, a flow inclined to the horizontal can be constructed by choosing complex C of correct magnitude and argument. Substituting $z = x + iy$ and $C = U_0 - iV_0$, we get the expressions for streamfunction (ψ) and potential function (ϕ) as

$$\phi(x, y) = U_0 x + V_0 y, \quad \psi(x, y) = U_0 y - V_0 x \tag{4.24}$$

The streamlines for different choices of C are shown in Fig. 4.1.

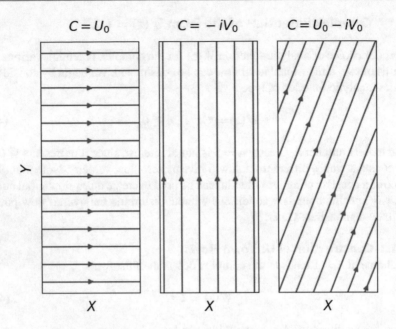

Fig. 4.1 Construction of uniform flow with complex potential $W(z) = Cz$. Direction of flow can be set by proper choice of constant C

4.1.1.2 Flow Inside a Wedge

Flow though a sector can be constructed by considering complex potential $W(z) = Cz^n$ with $n \geq 2$. Using polar representation of a complex number ($z = re^{i\theta}$), we get

$$\phi = Cr^n \cos n\theta, \quad \psi = Cr^n \sin n\theta \tag{4.25}$$

Here, C is considered to be a real number. Recall that a streamline is placed along a constant ψ contour across which there can not be any flow penetration. Especially, we look for $\psi = 0$ contour obtained for $\theta = 0$ and $\theta = \pi/n$. Therefore, if we keep a wedge-like obtain on $\psi = 0$ contour, the flow inside the wedge will remain undisturbed. Therefore, a flow inside a wedge can be represented by streamfunction and potential function given by Eq. (4.25). The streamlines are shown in Fig. 4.2. The components of the fluid velocity for this potential are given by

$$u = Cnr^{n-1} \left(\cos n\theta \cos \theta + \sin n\theta \sin \theta \right)$$

$$v = -Cnr^{n-1} \left(\sin n\theta \cos \theta - \cos n\theta \sin \theta \right)$$

Fig. 4.2 Streamlines for flow inside the wedge with complex potential $W(z) = Cz^2$ for $n \geq 2$. The wedge is placed on the $\psi = 0$ contour

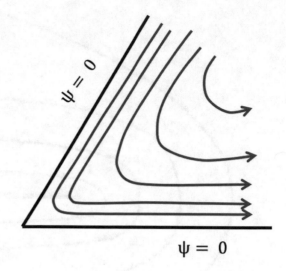

$\psi = 0$

Comparing with Eq. (4.19), the radial (u_r) and polar (u_θ) components of the velocity are given by

$$u_r = Cnr^{n-1} \cos n\theta, \quad u_\theta = -Cnr^{n-1} \sin n\theta. \tag{4.26}$$

Note that at $\theta = 0$, $u_r = Cnr^{n-1}$, $u_\theta = 0$ and for $\theta = \pi/n$, $u_r = -Cnr^{n-1}$, $u_\theta = 0$. Therefore, for positive (negative) values of C, the flow enters(leaves) radially from $\theta = \pi/n$ and leaves(enters) radially from $\theta = 0$. Thus, the magnitude and the direction of flow can be set by the proper choice of C.

4.1.1.3 Flow over a Sharp Edge

The flow over a sharp edge can be constructed with a complex potential of the form

$$W(z) = Cz^{1/2}, \tag{4.27}$$

where C is a real number. The corresponding potential and streamfunction are written as

$$\phi = Cr^{1/2} \cos \frac{\theta}{2}, \quad \psi = Cr^{1/2} \sin \frac{\theta}{2}. \tag{4.28}$$

Therefore, the sharp edge at $z = 0$ is represented by $\theta = 0$ and $\theta = 2\pi$ for which we get a constant streamfunction contour of $\psi = 0$. The streamlines associated with this flow are shown in Fig. 4.3. Note that the complex velocity is singular at the edge ($z = 0$) as it is inversely proportional to $\sqrt{r}$. In reality, the assumptions made in the potential flow theory are no longer valid in the region near singularity due to sharp rise in the fluid velocity.

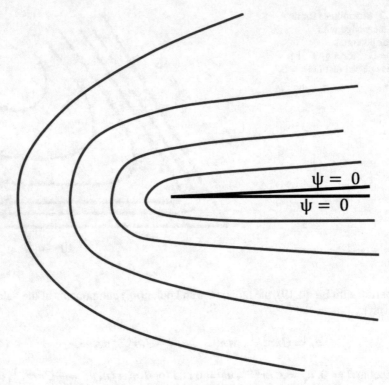

Fig. 4.3 Flow over a sharp edge represented by a complex potential of the form $W(z) = Cz^{1/2}$

4.1.1.4 Flow due to a Doublet

A doublet flow can be constructed with a complex potential

$$W(z) = \frac{C}{z} \tag{4.29}$$

Note that this function is analytic everywhere except $z = 0$. Therefore, in the region outside the singularity, this complex potential can represent an ideal fluid flow. The corresponding potential and streamfunction are expressed as

$$\phi = \frac{Cx}{x^2 + y^2}, \quad \psi = \frac{-Cy}{x^2 + y^2} \tag{4.30}$$

Therefore, the streamlines are in the form of a circle

$$x^2 + y^2 + \frac{C}{\psi}y = 0, \tag{4.31}$$

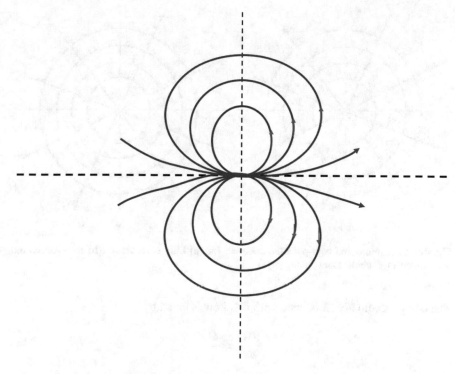

Fig. 4.4 Streamline for the doublet flow with complex potential $W(z) = C/z$. Such flow is obtained by superposition of line source and sink flows

whose radius is $C/2\psi$ with a centre at $(0, -C/2\psi)$. The streamlines associated with double are shown in Fig. 4.4. In the next section, we will demonstrate that a doublet flow, as the name suggests, is formed by the superposition of line source and sink.

4.1.2   Complex Potentials of the Form $W(z) = C \log z$

Now we consider a complex potential of the form

$$W(z) = C \log z. \tag{4.32}$$

Since $z = re^{i\theta} = Re^{i(\theta+2\pi)}$, we consider only the range $0 \le \theta < 2\pi$ so that the potential remains a single valued function. Separating into real and imaginary part we get

$$W(z) = C \log r + iC\theta. \tag{4.33}$$

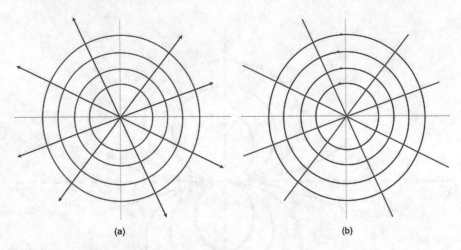

Fig. 4.5 Streamlines and equi-potential contours for (**a**) Line source/sink ((**b**) line vortex) are respectively rays (circles) and circles (rays)

The complex velocity associated with this flow is given by

$$\frac{dW}{dz} = \frac{C}{z}. \tag{4.34}$$

This velocity becomes singular at the origin. Identical flow shifted to an arbitrary location (z_0) is obtained by considering a complex potential of the form

$$W(z) = C \log (z - z_0). \tag{4.35}$$

4.1.2.1 Line Source/Sink Flows

For a real number $C = Q_0/2\pi$, the potential and stream functions take the form

$$\phi = \frac{Q_0}{2\pi} \log r, \quad \psi = \frac{Q_0}{2\pi}\theta. \tag{4.36}$$

This shows that the streamlines are along constant θ direction as shown in Fig. 4.5. Therefore, for $Q_0 > 0$ we get a flow moving radially outwards form the origin where for $Q_0 < 0$ the flow is radially inwards towards the origin. The complex velocity associated with this flow is given by

$$\frac{dW}{dz} = u - iv = \frac{Q_0}{2\pi r} \cos\theta - i\frac{Q_0}{2\pi r} \sin\theta. \tag{4.37}$$

Therefore, using Eq. (4.19) we verify the flow is purely in the radial direction as the corresponding r and θ velocity components are

$$u_r = \frac{Q_0}{2\pi r}, \quad u_\theta = 0. \tag{4.38}$$

Thus, Q_0 can be interpreted as the flow rate per unit area. Now we consider a line source and sink of each magnitude (Q_0) located at $z = \pm z_0$ respectively. The complex potential due to both of them can be written using super position principle as

$$W(z) = \frac{Q_0}{2\pi} \log(z - z_0) - \frac{Q_0}{2\pi} \log(z + z_0) = \frac{Q_0}{2\pi} \log\left(\frac{z - z_0}{z + z_0}\right) \tag{4.39}$$

In a far-field limit ($z_0/z \ll 1$), we can neglect the terms of the order $(z_0/z)^2$. This leads to approximate potential of the form similar to that of a doublet as

$$W(z) \simeq \frac{Q_0}{2\pi} \log\left[1 + \frac{2z_0}{z}\right] \simeq \frac{2Q_0 z_0}{2\pi z} \tag{4.40}$$

4.1.2.2 Line Vortex Flows

For an imaginary number $C = i\Gamma$, we get potential and streamfunction are of the form

$$\phi = -\Gamma\theta, \quad \psi = \Gamma \log r. \tag{4.41}$$

Therefore, streamlines are contours of constant R representing a vortex flow around the origin as shown in Fig. 4.5. The complex velocity is given by

$$\frac{dW}{dz} = u - iv = \frac{\Gamma}{r} \sin\theta + i\frac{\Gamma}{r} \cos\theta. \tag{4.42}$$

Using Eq. (4.19), we get velocity in the polar coordinate system as

$$u_r = 0, \quad u_\theta = -\frac{\Gamma}{r} \tag{4.43}$$

Thus, with a choice of sign of Γ one can set the direction of vortex flow.

4.1.3   Potential Flow Past a Cylinder

Since complex potential formed due to superposition of two or more analytic functions also represents a potential flow of ideal fluids, we consider a potential of the form

$$W(z) = U_0 z + \frac{C}{z}. \tag{4.44}$$

This potential represents a flow field due to combination of uniform flow in the horizontal direction and a doublet situated at the origin. The value of this potential on a circle of radius R_0 is given by substituting $z = R_0 e^{i\theta}$. Therefore,

$$W(z = R_0 e^{i\theta}) = \left(U_0 R_0 + \frac{C}{R_0} \right) \cos\theta + i \left(U_0 R_0 - \frac{C}{R_0} \right) \sin\theta.$$

Thus, the streamfunction on a circle of radius R_0, centred at the origin, is given by

$$\psi(z = R_0 e^{i\theta}) = \left(U_0 R_0 - \frac{C}{R_0} \right) \sin\theta. \tag{4.45}$$

For $C = U_0 R_0^2$, we get $\psi = 0$, on the circle. Since $u_r = (1/r)\partial\psi/\partial r$, the radial velocity of the fluid on the circumference of the circle is zero. In other words, the fluid flow outside the circle can not penetrate through the circle. Therefore, if we introduce an infinite cylinder of radius R_0 in place of this circle, outside flow (for $r > R_0$) given by this complex potential should remain unaltered by the introduction of this cylinder. Thus, the flow past a cylinder of radius R_0 is represented by a complex potential

$$W(z = r e^{i\theta}) = U_0 \left(z + \frac{R_0^2}{z} \right), \quad \forall \ r > R_0. \tag{4.46}$$

The corresponding streamfunction is expressed as

$$\psi(r, \theta) = U_0 r \left[1 - \left(\frac{R_0}{r} \right)^2 \right] \sin\theta, \quad \forall \ r > R_0. \tag{4.47}$$

Note that for $r \gg R_0$, we get the streamfunction corresponding to the horizontal flow whereas constant ψ contour is obtained on a circle of radius R_0. The corresponding flow pattern shown in Fig. 4.6 indeed mimics a uniform horizontal flow passing a cylinder of radius R_0.

Exercise 4.2 Show that the pressure field around the cylinder is given by the expression

$$p = p_\infty + \rho U_0^2 \left(\frac{R_0}{r} \right)^2 \left[\cos 2\theta - \frac{1}{2} \left(\frac{R_0}{r} \right)^2 \right] \tag{4.48}$$

where p_∞ is the pressure at $r = \infty$.

Fig. 4.6 Flow past a stationary cylinder obtained by superposition of uniform flow and a doublet. Strength of a doublet is adjusted to get $\psi = 0$ contour on the surface of the cylinder

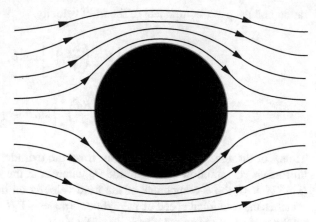

4.1.4   Flow Past a Rotating Cylinder

Flow past a rotating cylinder can be modelled by adding a line vortex at the origin to the complex potential associated with the fast past a stationary cylinder described earlier. Therefore, complex potential associated with such a flow is given by

$$W(z = re^{i\theta}) = U_0 \left(z + \frac{R_0^2}{z} \right) + \frac{i\Gamma}{2\pi} \log \left(\frac{z}{R_0} \right), \quad \forall \ r > R_0. \tag{4.49}$$

Separating real and imaginary parts of this complex function, we obtain the expressions for potential and streamfunction for the flow past a rotating cylinder as

$$\phi = U_0 r \left[1 + \left(\frac{R_0}{r} \right)^2 \right] \cos\theta - \frac{\Gamma}{2\pi} \theta$$

$$\psi = U_0 r \left[1 - \left(\frac{R_0}{r} \right)^2 \right] \sin\theta + \frac{\Gamma}{2\pi} \log \left(\frac{r}{R_0} \right) \tag{4.50}$$

Note that on the circle $z = R_0 e^{i\theta}$, we have $\psi = 0$ i.e. a constant ψ contour. Thus, similar to flow past a stationary cylinder case, our choice of complex potential has lead to a vanishing radial velocity (u_r) on the circle of radius R_0. This implies that this circle can mimic a rotating cylinder and the complex potential described by Eq. (4.49) represents a flow past such a cylinder. Using Eq. (4.14), we obtain the radial and the polar component of the fluid velocity.

$$u_r = U_0 \left[1 - \left(\frac{R_0}{r} \right)^2 \right] \cos \theta,$$

$$u_\theta = -U_0 \left[1 + \left(\frac{R_0}{r} \right)^2 \right] \sin \theta - \frac{\Gamma}{2\pi r} \qquad (4.51)$$

Thus, for distances considerably away from the cylinder ($r/R_0 \gg 1$) we get the uniform horizontal flow of magnitude U_0 whereas in the absence of incoming flow (i.e. $U_0 = 0$) the θ-component of the fluid velocity on the surface of the cylinder matches the rotational speed of the cylinder ($u_\theta = -\Gamma/(2\pi R_0)$, as expected. Also, on the surface of the cylinder ($z = R_0 e^{i\theta}$) we get

$$u_r(R_0, \theta) = 0, \quad u_\theta(R_0, \theta) = -\frac{\Gamma}{2\pi R_0} - 2U_0 \sin \theta \qquad (4.52)$$

Thus, the radial component is flow velocity is zero everywhere on the surface of the cylinder on the other hand, θ-component is non-zero. But there may exist a special point on the cylinder where flow velocity is identically zero. This point is called as the stagnation point. Equating $u_\theta(R_0, \theta_s)$ to zero, get the location of this point as

$$\sin \theta_s = -\frac{\Gamma}{4\pi U_0 R_0} \qquad (4.53)$$

This shows that for a stationary cylinder ($\Gamma = 0$), the stagnation point is located at angular positions $\theta_s = 0, \pi$ on the surface of the cylinder. When the cylinder's rotational speed ($\Gamma/2\pi R_0$) is less that $2U_0 R_0$ i.e. $\Gamma < 4\pi U_0 R_0$, two distinct stagnation points are present in the third and fourth quadrant for a positive value of Γ. Note that in these quadrants, the incoming flow and flow due to rotation oppose each other whereas in the first two quadrant they reinforce each other for $\Gamma > 0$. With increasing rotation speed (Γ) the two stagnation points move towards $\theta = 3\pi/2$ whereas they merge with each other for a critical value of $\Gamma = 4\pi U_0 R_0$. Finally, when the rotation speed is above this critical value ($\Gamma > 4\pi U_0 R_0$), we can not have a stagnation point on the surface of the cylinder as $\sin \theta_s$ must be less than unity. In such a case, the point moves out of the cylinder and the location of point can be found by look at solution of Eq. (4.51) for $R_0/r \neq 1$ (Fig. 4.7).

4.2   Conformal Transformations

We have seen how to construct elementary flow patterns for an ideal fluid using various complex potentials. We also demonstrated that more complicated flow patterns can be produced using superposition of two or more complex potentials.

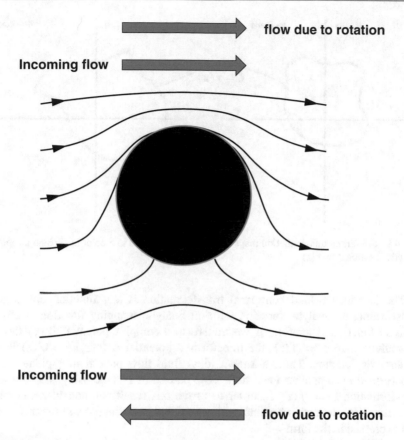

Fig. 4.7 Streamlines for the flow past a rotating cylinder. Note that the stagnation points lie in the third and fourth quadrant as in these quadrants flow due to rotation is in the opposite direction that of the incoming flow

For example, by adding complex potentials associated with uniform flow, a doublet and a line vortex we could generate a pattern for flow over a rotating cylinder. Apart from superposition principle, conformal transformation is another method to produce complicated flow patterns from known complex potentials associated with elementary flows.

In conformal transformation, we consider a mapping between z-plane to ζ-plane with an analytic function f such that

$$\zeta = f(z) = \xi + i\eta, \tag{4.54}$$

where ξ, η are the real and imaginary parts of ζ respectively. Geometrically, this transformation maps every point (x, y) in z-plane into a point (ξ, η) in ζ-plane. Therefore, a boundary of known shape in one complex plane can transformed into arbitrarily shaped boundary in another complex plane (Fig. 4.8).

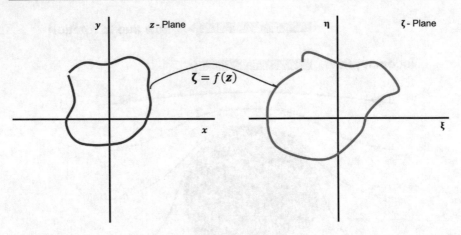

Fig. 4.8 Conformal transformation maps every point in z-plane to a point in ζ-plane through an analytic function $\zeta = f(z)$

The key idea behind conformal transformation is any function obtained by transforming an analytic function with an analytic mapping function is also an analytic function. Therefore, if we transform a complex potential $W(\zeta)$ through an analytic map $\zeta = f(z)$, the transformed potential $W(\zeta(z)) = W(z)$ is also an analytic function. Thus, a known ideal fluid flow pattern in ζ-plane can be transformed in to another ideal fluid flow pattern in z-plane using the conformal transformation $\zeta = f(z)$. In order to prove this result, we consider a complex potential $W(\zeta)$, defined in ζ-plane, with associated potential (ϕ) and streamfunction (ψ) expressed in the form

$$W(\zeta) = \phi(\xi, \eta) + i\psi(\xi, \eta). \tag{4.55}$$

Since $W(\zeta)$ is an analytic function, ϕ, ψ must satisfy Cauchy-Riemann relations

$$\frac{\partial \phi}{\partial \xi} = \frac{\partial \psi}{\partial \eta}, \quad \frac{\partial \phi}{\partial \eta} = -\frac{\partial \psi}{\partial \xi} \tag{4.56}$$

Also, on account of $f(z) = \zeta = \xi(x, y) + i\eta(x, y)$ being analytic function Cauchy-Riemann relations are expressed as

$$\frac{\partial \xi}{\partial x} = \frac{\partial \eta}{\partial y}, \quad \frac{\partial \xi}{\partial y} = -\frac{\partial \eta}{\partial x}. \tag{4.57}$$

Using the transformation, we can express both potential and streamfunctions in the z-plane i.e.

$$\phi(\xi, \eta) \equiv \phi(\xi(x, y), \eta(x, y)), \quad \psi(\xi, \eta) \equiv \psi(\xi(x, y), \eta(x, y))$$

Therefore, using chain-rule we write

$$\frac{\partial \phi}{\partial x} = \frac{\partial \phi}{\partial \xi}\frac{\partial \xi}{\partial x} + \frac{\partial \phi}{\partial \eta}\frac{\partial \eta}{\partial x},$$

$$\frac{\partial \phi}{\partial y} = \frac{\partial \phi}{\partial \xi}\frac{\partial \xi}{\partial y} + \frac{\partial \phi}{\partial \eta}\frac{\partial \eta}{\partial y},$$

$$\frac{\partial \psi}{\partial x} = \frac{\partial \psi}{\partial \xi}\frac{\partial \xi}{\partial x} + \frac{\partial \psi}{\partial \eta}\frac{\partial \eta}{\partial x},$$

$$\frac{\partial \psi}{\partial y} = \frac{\partial \psi}{\partial \xi}\frac{\partial \xi}{\partial y} + \frac{\partial \psi}{\partial \eta}\frac{\partial \eta}{\partial y}. \tag{4.58}$$

Substituting Eqs. (4.56) and (4.57) in above relations, we see that Cauchy-Riemann relations for ϕ and ψ are also satisfied in the z-plane as well, i.e.

$$\frac{\partial \phi}{\partial x} = \frac{\partial \psi}{\partial y}, \quad \frac{\partial \phi}{\partial y} = -\frac{\partial \psi}{\partial x} \tag{4.59}$$

Therefore, the transformed complex potential $W(z) = \phi(x, y) + i\psi(x, y)$ is also an analytic function which represents an ideal fluid flow. This also implies that if any function satisfies the Laplace's equation in ζ-plane then the conformation transformation ensures that the transformed function is also satisfies Laplace's equation in the z-plane.

Since we have established that a complex potential from one plane is transformed into a complex potential in another plane, we now check how the two complex corresponding velocities are related. Let $\mathbf{U}(z)$ and $\mathbf{U}(\zeta)$ be the complex velocities, associated with complex potentials $W(z)$ and $W(\zeta)$ in z and ζ-plane respectively. Using chain rule we see that

$$\mathbf{U}(z) = \frac{dW(z)}{dz} = \frac{dW(\zeta)}{d\zeta}\frac{d\zeta}{dz}.$$

Using $\mathbf{U}(\zeta) = dW(\zeta)/d\zeta$ and $f'(z) = d\zeta/dz$, we obtain the relation between two complex velocities as

$$\mathbf{U}(z) = f'(z)\mathbf{U}(\zeta) \tag{4.60}$$

Thus, complex velocities are mapped into each other through a derivative of the mapping function. Clearly the mapping will fail when $f'(z)$ becomes zero or infinity in which case one-to-one correspondence between complex velocities in the two plane is lost. Also, note that in general the two velocities are proportional to each other. Due to this proportionality constant, the angle between two velocity vectors which are passing through a common point is preserved in both planes. In order to demonstrate the application of conformal mapping, we consider a simple uniform horizontal plane in the ζ-plane whose complex potential is written as

$$W(\zeta) = U_0\zeta = U_0\xi + iU_0\eta. \tag{4.61}$$

Therefore, the potential and stream function contours are vertical and horizontal lines in ζ plane as shown in the figure. Now we apply as conformal transformation, $\eta = f(z) = z^2$. Then the transformed complex potential in the z-plane is obtained as

$$W(z) = U_0z^2 = U_0(x^2 - y^2) + i(2U_0xy), \tag{4.62}$$

with associated potential function $\phi = U_0(x^2 - y^2)$ and stream function $\psi = 2U_0xy$. Thus, uniform flow in the ζ-plane is transformed into a flow between perpendicular wall as shown in the figure below.

4.2.1 Joukowski Transformation

One of the most important examples of conformal transformation is the Joukowski transformation. In this transformation, every point in ζ-plane is mapped into z-plane through a mapping function

$$z = \zeta + \frac{c^2}{\zeta}. \tag{4.63}$$

Typically, c^2 is a real number. With this transformation, the complex velocities in the two planes are related through

$$\mathbf{U}(\zeta) = \frac{dz}{d\zeta}\mathbf{U}(z) = \left(1 - \frac{c^2}{\zeta^2}\right)\mathbf{U}(z). \tag{4.64}$$

Thus, for large values of $|\zeta|$ we see that the transformation leads to identity relation of the type $z \to \zeta$. Consequently the velocities in the z and ζ planes become equal as we move sufficient distance away from the origin, more precisely when $c^2/\zeta^2 \to 0$. In other words, if the body is placed around origin then the incoming and outgoing flows towards/away from the body are identical and difference in the flow field due to shape of the body can be seen in the region in the origin. Also, since the body is placed around origin ($\zeta = 0$) the associated singularity in the mapping function as well as the complex velocity field is not included in the flow field. Therefore, similar to flow past cylinder case, this singularity is of no physical consequence. Note that both points $\zeta = +c$ and $\zeta = -c$ are mapped to a same point in z-plane for which velocity mapping between two planes is not unique. Again, such a situation is avoided by considering the shape of the body in ζ-plane such that the critical points $\zeta = \pm c$ do not lie inside the flow field.

4.2.2 Joukowski Transformation for Flow Past an Ellipse

Consider a circle of radius R_0 centred around origin in the ζ-plane such that $R_0 > c$ as shown in the figure with $c^2 > 0$. Using Joukowski transformation given by Eq. (4.63), every point on this circle is mapped onto a point in the z-plane. The locus of these points can be found by substituting $\zeta = R_0 e^{i\theta}$, i.e.

$$z = x + iy = R_0 e^{i\theta} + \frac{c^2}{R_0} e^{-i\theta}. \tag{4.65}$$

Separating real and imaginary parts, we get

$$x = \left(R_0 + \frac{c^2}{R_0} \right) \cos\theta,$$

$$y = \left(R_0 - \frac{c^2}{R_0} \right) \sin\theta. \tag{4.66}$$

Therefore, points on the circle $\zeta = R_0 e^{i\theta}$ in ζ-plane are mapped onto locus of points in z-plane with equation

$$\frac{x^2}{\left(R_0 + \frac{c^2}{R_0} \right)^2} + \frac{y^2}{\left(R_0 - \frac{c^2}{R_0} \right)^2} = 1. \tag{4.67}$$

This is an equation of an ellipse with semi-major axis (along x-direction) of length $R_0 + c^2/R_0$ and semi-minor axis (along y-direction) of length $R_0 - c^2/R_0$. Thus, a circle in ζ-plane is mapped onto an ellipse into the z-plane. Note that for $c^2 < 0$ the major and minor axes get flipped and the ellipse gets rotated by 90° compared to $c^2 > 0$ case as shown in the figure.

Now, a uniform flow passing over a circular cylinder of radius R_0 in ζ-plane is given by

$$W(\zeta) = U_0 \left(\zeta + \frac{R_0^2}{\zeta} \right) \tag{4.68}$$

The corresponding transformed potential in the z-plane after substituting the inverse Joukowski mapping $\zeta = (z/2) \pm \sqrt{(z/2)^2 - c^2}$ is given by

$$W(z) = U_0 \left(\frac{z}{2} + \sqrt{\left(\frac{z}{2} \right)^2 - c^2} + \frac{R_0^2}{\frac{z}{2} + \sqrt{\left(\frac{z}{2} \right)^2 - c^2}} \right) \tag{4.69}$$

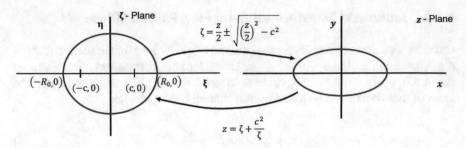

Fig. 4.9 A circle in ζ-plane is transformed into an ellipse in z-plane using Joukowski mapping

This can be further simplified into

$$W(z) = U_0 \left[z + \left(\frac{R_0^2}{c^2} - 1 \right) \left(\frac{z}{2} - \sqrt{\left(\frac{z}{2} \right)^2 - c^2} \right) \right]. \tag{4.70}$$

Note that at large distances from the origin ($(z/2)^2 \gg c^2$) the complex potential tends to $U_0 z$, consistent with the uniform incoming flow. On the other hand, the second term represents the deviations in the flow pattern due to elliptic shape of the body placed around the origin. Substituting $z = x + iy$ and separating real and imaginary part of the complex function we can calculate the corresponding potential and streamfunction. The stagnation points on this ellipse can be calculated by using the mapping function on the stagnation points on a circular cylinder in the ζ-plane (Fig. 4.9).

4.2.3 Joukowski Transformation for Flow over an Airfoil

We now demonstrate application of Joukowski transformation to construct a flow over an symmetric airfoil. For construction of an ellipse in z-plane we considered a circle of radius R_0 which is centred at origin in ζ-plane with $R_0 > c$. This resulted in squeezing of circle along one axis leading to the formation of an ellipse. If we consider a circle passing through points $(c, 0)$ and $(-c, 0)$ such that $R_0 = |c|$, the circle gets transformed into a flat-plate as can be seen from Eq. (4.66) with $x = 2R_0 \cos \theta$ and $y = 0$. Thus, when the circle passes through points $(\pm c, 0)$ a sharp-edge is created at the corresponding points in the z-plane. For the case of a symmetric airfoil, we consider an intermediate case between an ellipse and a flat plate.

For a symmetry airfoil, we consider a circle of radius R_0, centred at $(-b, 0)$, passing through a point $(c, 0)$ as shown in the figure. This condition requires

$$R_0 = b + c$$

Now, a typical point on this circle in the ζ-plane is represented by $R(\theta)e^{i\theta}$ as shown in the figure. Therefore, the radius (R_0) satisfies the relationship

$$R_0^2 = (R\cos\theta + b)^2 + (R\sin\theta)^2 = R^2 + b^2 + 2Rb\cos\theta.$$

Using $R_0 = b + c$, we get the quadratic equation for R as

$$R^2 + 2Rb\cos\theta - (c^2 + 2bc) = 0,$$

whose solution is written in the form

$$R = -b\cos\theta \pm \sqrt{b^2\cos^2\theta + c^2 + 2bc}$$

Taking positive root since R must be positive, we get

$$R = c\left[\sqrt{1 + \frac{2b}{c} + \frac{b^2}{c^2}\cos\theta} - \frac{b}{c}\cos\theta\right] \tag{4.71}$$

Having succeeded in expressing every point on the circle in terms of b and c, we can now use Joukowski mapping to compute the transformed curve in z-plane. In particular, we look for the case when $b/c \ll 1$ i.e. the circle is slightly displaced from the origin. In this case,

$$R \simeq c + b\,(1 - \cos\theta) \tag{4.72}$$

The transformed curve in the z-plane using Joukowski mapping is expressed as

$$z = c\left[1 + \frac{b}{c}(1 - \cos\theta)\right]e^{i\theta} + \frac{ce^{-i\theta}}{1 + \frac{b}{c}(1 - \cos\theta)} \tag{4.73}$$

Again in $b/c \ll 1$ limit, we get

$$z \simeq c\left[1 + \frac{b}{c}(1 - \cos\theta)\right]e^{i\theta} + c\left[1 - \frac{b}{c}(1 - \cos\theta)\right]e^{-i\theta}$$

Separating real and imaginary parts, the transformed coordinate in the z-plane is obtained through components a

$$x \simeq 2c\cos\theta, \quad y \simeq 2b\,(1 - \cos\theta)\sin\theta \tag{4.74}$$

Eliminating θ from above equation, we show that the circle in the ζ-plane is transformed into a curve in the z-plane whose equation is of the form

$$y \simeq \pm 2b \left(1 - \frac{x}{2c}\right) \left(1 - \frac{x^2}{4c^2}\right)^{1/2}. \tag{4.75}$$

This is the equation of symmetric airfoil when $b \ll c$. It is easy to show that the parameters b and c of the mapping can be expressed in terms of length (l) and maximum thickness (t) of the airfoil as

$$c = \frac{l}{4}, \quad b = \frac{t}{3\sqrt{3}} \tag{4.76}$$

Therefore, the complex potential for the flow past a symmetric airfoil of length l and maximum thickness t is given by the expression

$$W(z) = U_0 \left[\frac{z}{2} + \sqrt{\left(\frac{z}{2}\right)^2 - \frac{l^2}{16}} + \frac{\left(\frac{l}{4} + \frac{t}{3\sqrt{3}}\right)^2}{\frac{z}{2} + \sqrt{\left(\frac{z}{2}\right)^2 - \frac{l^2}{16}}} \right] \tag{4.77}$$

Note that this expression is obtained by substituting expression in Eq. (4.76) in Eq. (4.69). Finally, separating real and imaginary parts of the complex potential we can compute the potential and streamfunctions for the ideal flow passing over a symmetric airfoil. In fact, similar procedure can be repeated for computing flow over different types of airfoils typically referred 'family of Joukowski airfoils' (Fig. 4.10).

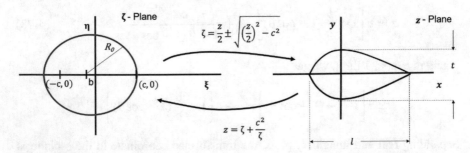

Fig. 4.10 Joukowski transformation for mapping a circle into a symmetric airfoil of length l and maximum thickness t. The choice mapping parameters c, b, R_0 determine the thickness and length of the airfoil

4.3 Force on a Body Immerse in a Potential Flow

Having derived the expressions for streamfunction/potential function for flow past bodies of various shapes, we now focus on calculation of force experienced by such objects. We will derive the expression for the force on a two dimensional body of arbitrary shape immersed inside the potential flow. The key idea behind these calculations comes from the fact that the force experienced by the body is primarily the reaction force of the fluid which is undergoing a rate of change of momentum due to the presence of the body. Therefore, when the fluid is deflected by the presence of body its momentum changes causing equivalent force in the opposite direction on the body.

Let's a consider a arbitrary contour C around the body immersed inside an ideal fluid. By definition, the body forms a streamline as we have shown in the previous section. Therefore the fluid can't pass through the body and the net rate of change of moment density for the fluid between the body and the contour C is given by

$$\frac{d\mathbf{P}}{dt} = \int_C \rho \mathbf{u}(u\,dy - v\,dx) \tag{4.78}$$

Note that $\rho\,(u\,dy - v\,dx)$ represents the mass flux through the infinitesimal differential element of C. In steady state, this rate of change of momentum must be balanced by the pressure force along C and the net force exerted by the body on the fluid which is the negative of the force experience by the body ($\mathbf{F}$). The pressure force on the differential element of C is given by $-p(dy\hat{x} - dx\hat{y})$. Therefore, the x and Y components of the force on the body are given by

$$F_x = -\int_C p\,dy - \int_C \rho u(u\,dy - v\,dx) \tag{4.79}$$

$$F_y = \int_C p\,dx - \int_C \rho v(u\,dy - v\,dx) \tag{4.80}$$

Using the Bernoulli's equation for pressure i.e. $P + \rho(u^2 + v^2)/2 = \text{constant}$, one can show that the forces can be expressed in terms of complex velocity $(u - iv)$ in the following form:

$$F_x - iF_y = i\frac{\rho}{2}\int_C (u - iv)^2\,dz \tag{4.81}$$

where $dz = dx + idy$. This equation, famously known as Blasius integral law, allows us to compute the components of the force experienced by the body especially when the potential flow solution is available.

Exercise 4.3 Starting with Eqs. (4.79) and (4.80), prove the result of Eq. (4.81).

4.3.1   Force on Cylinder with Circulation Around It

Now, we use the Blasius integral law, to compute the force on a circular cylinder. A uniform horizontal flow of magnitude U_0 is passing over this cylinder. Additionally, a circulation Γ is also set around it. The complex potential associated with this flow is given by

$$W(z) = U_0 \left(z + \frac{R_0^2}{z} \right) + \frac{i\Gamma}{2\pi} \log \left(\frac{z}{R_0} \right) \qquad (4.82)$$

The corresponding complex velocity can be expressed as

$$u - iv = \frac{dW}{dz} = U_0 \left(1 - \frac{R_0^2}{z^2} \right) + \frac{i\Gamma}{2\pi z} \qquad (4.83)$$

Substituting above expression in the Blasius integral law, we get

$$F_x - iF_y = i\frac{\rho}{2} \int_C \left[U_0^2 + \frac{iU_0\Gamma}{\pi z} - \frac{2U_0^2 R_0^2}{z^2} - \frac{\Gamma^2}{4\pi^2 z^2} - \frac{iU_0\Gamma R_0^2}{\pi z^3} + \frac{U_0^2 R_0^4}{z^4} \right] dz$$
$$(4.84)$$

Applying residue theorem to this complex contour integral, we get

$$F_x - iF_y = -i\rho U_0 \Gamma \qquad (4.85)$$

Note that F_x is the force experienced by the body in the direction of the incoming flow and in the context of aerodynamics it is typically called as a drag force. On the other hand, F_y represents the force acting perpendicular to the incoming flow direction and is named as lift force. Our calculations show that the circular cylinder experiences finite lift force in the presence of finite circulation around it. It is this circulation around the wings of an aircraft which makes its flight possible !!

4.3.2   D'Alembert's Paradox and Its Resolution

The calculations of hydrodynamic force show that a circular cylinder immersed in an ideal (inviscid and incompressible) fluid experiences zero drag force but finite lift force in the presence of circulation around it. In fact, this result can be generalized to an arbitrarily shaped body. The conclusion of zero drag force on a body is in direct contradiction with the everyday observation of a body experiencing a resistance while moving inside the fluid. This paradox was first pointed by D'Alembert in 1749.

The solution to this paradox lies in considering the finite viscosity of the fluids as against the inviscid fluid approximation considered in the potential flow theory. In fact, Stokes in 1851 derived his famous formula for calculating the drag force on a sphere by considering the highly viscous fluid. The issue was still not resolved for high speed flows in standard fluids such as air and water where potential flow theory was remarkably successful. For such flows, prima-facie the neglect of viscous effects looked fairly justifiable. This can be seen by comparing the inertial term ($\rho \mathbf{u} \cdot \nabla \mathbf{u}$) with the viscous term ($\mu \nabla^2 \mathbf{u}$) in the incompressible Navier-Stokes equation. For high speed flows, the inertial terms dominates over the viscous term in bulk of the fluid as the former terms increases with square of the velocity ($\sim u^2$) whereas the later diminishes with the square of the distance ($\sim 1/L^2$). Therefore, neglecting viscous term, it was believed that one can safely use potential flow theory to describe the high speed flow. But D'Alembert's paradox clearly demonstrated inconsistency with this argument.

Finally, the paradox was resolved by Prandtl in 1904 by suggesting that although viscous term is negligible in the bulk of the fluid, it can be comparable to inertial term in a region very near to the body. This narrow region around the body is called as 'boundary layer'. Inside the boundary layer, the fluid velocity must change from zero (no-slip boundary condition) to incoming free-streaming velocity set by potential flow theory. Therefore, $\nabla^2 \mathbf{u}$ term can be large, especially in the direction normal to the surface. In other words, large change in the velocity inside the boundary layer results in non-negligible viscous dissipation eventually leading to the generation of drag. Note that in potential flow theory we always assumed that the surface of the body is defined by $\psi =$ constant i.e. a streamline. This, by definition, lead to a 'no-penetration' boundary condition i.e. the fluid velocity normal to the surface was set to zero. But the velocity along the surface was kept unspecified in the potential flow theory. This is mainly due to the fact that in the absence of viscous term, the Navier-Stokes equation becomes first order partial differential equation for which one boundary condition on the surface is sufficient. On the other hand, introduction of viscosity term allows us to specify all components of velocity. Therefore, inside the boundary layer the fluid velocity changes significantly so that the contribution of viscous term becomes non-negligible. Once these contributions are taken into account, the D'Alembert's paradox is resolved.

We can make simple estimate about the thickness of the boundary by assuming the flow to be laminar. Let's consider an incoming flow of typical magnitude U_0 over an object of size L. Because of the viscosity, the momentum will diffuse perpendicular to the surface of the object due to velocity gradients. We can then estimate the boundary layer thickness by simply evaluating how far the momentum will diffuse in the time required for flow to cross the object. This time (τ) will scale as

$$\tau \sim \frac{L}{U_0}$$

Considering v to be the kinematic viscosity (μ/ρ) of the fluid and noting that this viscosity has the dimensions of diffusivity (length^2/time), we can estimate the boundary thickness (δ) to be

$$\delta \sim \sqrt{v\tau} \sim \sqrt{\frac{vL}{U_0}}$$

With more precise calculation, one can show that the boundary thickness varies over the stream-wise length (x) of the body as

$$\delta(x) \simeq 5\sqrt{\frac{vx}{U_0}} \tag{4.86}$$

In reality, the presence of strong velocity shear ($\partial\mathbf{u}/dy$) makes boundary layer unstable leading formation of turbulent boundary layer. In this situation, the thickness gets altered significantly. In order to understand this dynamics, we need to study physics of turbulence which will be introduced in Chap. 14.

Another important consequence of formation of boundary layer around the body is the generation of vorticity inside this region. Considering the incoming flow in the horizontal direction, the velocity gradient in the boundary layer will be in the y-direction with a magnitude $\sim \partial u/\partial y$. This term will contribute to the generation of vorticity as the z-component of vorticity will be given by $\omega_z = \partial v/\partial x - \partial u/\partial y$. Integrating this vorticity component over the contour passing through the edge of the boundary layer will give us the circulation (Γ) around the body. Note that this was put in an adhoc way in the potential flow theory, but introduction of viscous boundary layer allows us to have self-consistent treatment of circulation. This of course comes at the cost of increased complexity in solving the Navier-Stokes equation, both analytically and numerically. Therefore, in spite of its limitations the potential theory remained one of most successful theory in the early development of aerodynamics.

Exercise 4.4 Starting with two dimensional Navier-Stokes equations for an incompressible fluid, show that the pressure across the boundary layer is constant. Assume that the typical thickness of the boundary layer (δ) is much smaller than the streamwise length of the body (x) i.e $\delta/x \ll 1$.

Viscous Flow

5

> *Everyone goes with the flow ... but the one who goes against the flow becomes someone remarkable in life.*
>
> —*Swami Vivekananda*

5.1 Poiseuille Flow

Consider a steady unidirectional flow of a viscous incompressible fluid with velocity field $\boldsymbol{u} = (u, 0, 0)$. Assume ρ, μ are constants, with gravity as the body force. The equation for $\boldsymbol{u}$ is

$$\rho \frac{d\boldsymbol{u}}{dt} = -\nabla p + \rho \mathbf{g} + \mu \nabla^2 \boldsymbol{u}. \tag{5.1}$$

For steady flow, $\partial \boldsymbol{u}/\partial t$ is zero whereas the term $\mathbf{u} \cdot \nabla \mathbf{u}$ is also identically zero if we consider length along x-direction of pipe carrying the fluid to be significantly greater than the other two directions ($\partial/\partial x \equiv 0$). For an incompressible fluid, (5.1) is then

$$0 = -\frac{\partial(p - \rho g x)}{\partial x} - \mu \left(\frac{\partial^2 u}{\partial y^2} + \frac{\partial^2 u}{\partial z^2} \right),$$

$$0 = -\frac{\partial(p - \rho g y)}{\partial y},$$

$$0 = -\frac{\partial(p - \rho g z)}{\partial z}, \tag{5.2}$$

S. R. Jain et al., *A Primer on Fluid Mechanics with Applications*,
https://doi.org/10.1007/978-3-031-20487-6_5

Define $\overline{p} = p - \rho \mathbf{g}.\mathbf{r}$. We consider a situation where p is only dependent on x, thus we have a simple equation:

$$\frac{\partial \overline{p}}{\partial x} = \mu \left(\frac{\partial^2 u}{\partial y^2} + \frac{\partial^2 u}{\partial z^2} \right). \tag{5.3}$$

Here the LHS is a function of x alone, and RHS of y, z. Separating variables with a constant G, we have

$$\frac{\partial \overline{p}}{\partial x} = -G, \qquad \frac{\partial^2 u}{\partial y^2} + \frac{\partial^2 u}{\partial z^2} = -G/\mu. \tag{5.4}$$

We solve the equation for u in cylindrical geometry,

$$\frac{d^2 u}{dr^2} + \frac{1}{r} \frac{du}{dr} = -G/\mu \tag{5.5}$$

with boundary condition, $u = 0$ at the radius a. The general solution of

$$\frac{1}{r} \frac{d}{dr} \left(r \frac{du}{dr} \right) = -\frac{G}{\mu} \tag{5.6}$$

is

$$u(r) = -\frac{G}{4\mu} r^2 + A \log r + B. \tag{5.7}$$

For u to be finite at $r = 0$, we must set A to zero. At $r = a$, $u = 0$, hence $B = Ga^2/4\mu$. The solution is [9]

$$u(r) = -\frac{G}{4\mu} (a^2 - r^2). \tag{5.8}$$

The flux can be easily calculated:

$$\Phi = \int_0^a u.d\mathbf{S}$$

$$= \frac{-Ga^4\pi}{8\mu}. \tag{5.9}$$

Exercise 5.1 Solve the Poiseuille flow problem for a pipe with elliptic cross-section by applying the condition $u(y, z) = 0$ on the pipe boundary, given by the equation $y^2/a^2 + z^2/b^2 = 1$.

5.2   Flow of Tar Down an Inclined Plane

We describe the flow of tar as it is poured in a steady state, down an inclined road, making an angle ψ with the local horizontal. The atmospheric pressure p is constant and we assume that the tangential stress vanishes. That is, at $y = 0$, $u = 0$, and, at $y = d$, $du/dy = 0$. The equation of motion is

$$\mu \frac{d^2u}{dy^2} = -G = -\rho g \sin \alpha. \tag{5.10}$$

Integrating once,

$$\mu \frac{du}{dy} = -\rho g \sin \alpha y + c_1.. \tag{5.11}$$

The boundary condition at $y = d$ gives $c_1 = \rho g d \sin \alpha$. Equation (5.11) becomes

$$\mu \frac{du}{dy} = \rho g (d - y) \sin \alpha. \tag{5.12}$$

Integrating once again,

$$\mu u = -\frac{\rho g (d - y)^2 \sin \alpha}{2} + c_2. \tag{5.13}$$

The boundary condition at $y = 0$ gives $c_2 = \rho g d^2 \sin \alpha / 2$. Thus,

$$u(y) = \frac{\rho g \sin \alpha}{2\mu} (2dy - y^2). \tag{5.14}$$

Flux per unit width is

$$\Phi = \int_0^d u \, dy = \frac{\rho g \sin \alpha d^3}{3\mu}. \tag{5.15}$$

Inverting it, we get the width

$$d = \left(\frac{3\mu \Phi}{\rho g \sin \alpha} \right)^{1/3}. \tag{5.16}$$

5.3   Stokes Problems

5.3.1   Stokes First Problem

Consider an incompressible viscous fluid resting on a horizontal plate. The extent of the plate and the fluid is considered to be infinite such that all fluid quantities are approximately uniform in the horizontal plane. This is certainly a valid approximation if we are interested in the flow sufficiently away from the edges of the plate. Assuming the plate to be in the $x - z$ plane passing through the origin, the infinite plate approximation allows us to set all derivatives of fluid quantities in the x and z directions to zero i.e. $\partial/\partial x \equiv \partial/\partial z \equiv 0$. Similarly, the fluid is considered to be infinite in the vertical direction (y) implying that we are interested in the flow field which is sufficiently away from the top surface of the fluid. At time $t = 0$ an impulse is applied to the plate such that the plate starts moving uniformly in the horizontal plane with a speed U_0 for $t > 0$. This initial set-up is known as Stokes first problem where we are interested in how fluid quantities vary with time and space under this scenario [109]. Choosing x axis along the motion of the plate, we can reduce the problem to two-dimensions ($x - y$). Neglecting gravity we write the governing equations in the following form:

$$\frac{\partial u}{\partial x} + \frac{\partial v}{\partial y} = 0,$$

$$\frac{\partial u}{\partial t} + u\frac{\partial u}{\partial x} + v\frac{\partial u}{\partial y} = -\frac{1}{\rho}\frac{\partial p}{\partial x} + v\left(\frac{\partial^2 u}{\partial x^2} + \frac{\partial^2 u}{\partial y^2}\right),$$

$$\frac{\partial v}{\partial t} + u\frac{\partial v}{\partial x} + v\frac{\partial v}{\partial y} = -\frac{1}{\rho}\frac{\partial p}{\partial y} + v\left(\frac{\partial^2 v}{\partial x^2} + \frac{\partial^2 v}{\partial y^2}\right). \tag{5.17}$$

Subject to boundary conditions on the velocity field in the form

$$u(x, y = 0, t) = U_0, \quad v(x, y = 0, t) = 0,$$

$$u(x, y \to \infty, t) = 0, \quad v(x, y \to \infty, t) = 0. \tag{5.18}$$

Using the fact that $\partial/\partial x \equiv 0$ the continuity equation the continuity equation reduces to

$$\frac{\partial v}{\partial y} = 0.$$

Therefore, using the boundary condition $v(y = 0) = 0$ at $y = 0$ we can write the solution of y-component of the velocity field as

$$v(x, y, t) = 0. \tag{5.19}$$

Substituting v in the y-component of momentum equation, we see that the pressure is uniform across the fluid i.e.

$$p(x, y, t) = P_0, \tag{5.20}$$

where P_0 is a constant. Finally, substituting solutions of v and p into the x-component of the momentum equation and setting $\partial u / \partial x = 0$ we get the diffusion equation for the x-component of the velocity field.

$$\frac{\partial u}{\partial t} = v \frac{\partial^2 u}{\partial y^2} \tag{5.21}$$

The parabolic partial differential equation (PDE) of the above type is typically converted into an ordinary differential equation (ODE) by introducing a non-dimensional self-similar variable

$$\eta = \frac{y}{\sqrt{vt}}. \tag{5.22}$$

Note that the choice of η is motivated by the fact that in a diffusion problem the diffusivity (η) sets up the natural length-scale of diffusion. For example, a typical distance a particle will travel through pure diffusion depends upon time (t) for which it is diffusing and can be estimated as $\sqrt{vt}$. Therefore, we seek the solution of Eq. (5.21) in the following form:

$$u = U_0 f(\eta). \tag{5.23}$$

With this substitution, the Eq. (5.21) reduces to an ODE of the form

$$\frac{d^2 f}{d\eta^2} + \frac{\eta}{2} \frac{df}{d\eta} = 0. \tag{5.24}$$

The corresponding boundary conditions, consistent with Eq. (5.18), are expressed as

$$f(\eta = 0) = 1, \quad f(\eta \to \infty) = 0 \tag{5.25}$$

The Eq. (5.24) can be easily integrated to give

$$f(\eta) = A \int_0^\eta e^{-\eta^2/4} d\eta + B \tag{5.26}$$

The constants of integration can be calculated from boundary conditions (Eq. (5.25)) as $A = -1/\sqrt{\pi}$ and $B = 1$. Therefore, the solution of Eq. (5.24) is written as

$$f(\eta) = 1 - \frac{2}{\sqrt{\pi}} \int_0^{\eta/2} e^{-\zeta^2} d\zeta = 1 - \text{erf}(\eta/2), \tag{5.27}$$

where the error function $\text{erf}(x)$ is defined as

$$\text{erf}(x) = \frac{2}{\sqrt{\pi}} \int_0^x e^{-\zeta^2} d\zeta. \tag{5.28}$$

Substituting the solution in the Eq. (5.23) we get the expression for the horizontal component of fluid velocity inside the fluid as

$$u = U_0 \left[ ]1 - \text{erf}\left(\frac{y}{2\sqrt{\nu t}} \right) \right] \tag{5.29}$$

5.3.2   Stokes Second Problem

The initial set-up of Stokes second problem is identical to that of the first problem where an incompressible fluid is placed on top of a infinite horizontal plate in the $x - z$ plane. In this problem, instead of uniformly moving plate, we have an oscillating plate with a frequency ω. Accordingly, the boundary condition on the $x-$ component of velocity field gets modified to

$$u(x, y = 0, t) = U_0 \cos \omega t \tag{5.30}$$

All other boundary conditions remain identical to Eq. (5.18). The oscillating plate introduces natural time-scale for the fluid flows i.e. $\tau = \omega^{-1}$. In Stokes first problem the timescale was set purely by the kinematic viscosity (ν). In contrast, in the second problem we expect that fluid flow to be influence by the interplay between τ and viscous timescale (ν^{-1}). Starting with two dimensional incompressible Navier-Stokes equation and following similar steps as the first problem, we get identical solutions for y-component of velocity ($v(x, y, t) = 0$) and pressure ($p(x, y, t) = P_0$). Also, the x-component of velocity is governed by the diffusion equation

$$\frac{\partial u}{\partial t} = \nu \frac{\partial^2 u}{\partial y^2}$$

In order to solve this equation, we seek the solution in the form

$$u(x, y, t) = \text{Re}\left[u_0(y) e^{i\omega t} \right]. \tag{5.31}$$

Note that this solution is motivated by the presence of natural timescale set-up by the frequency of oscillation (ω) of the plate. On the contrary, the self-similar solution saught in the Stokes first problem was inspired from the observation that the timescale of viscous dissipation depends upon both kinematic viscosity and the distance up to which the dissipation occurs. Substituting Eq. (5.31) into the diffusion equation we get

$$\frac{d^2 u_0}{dy^2} = \frac{i\omega}{\nu} u_0. \tag{5.32}$$

Therefore, using $\sqrt{i} = (1+i)/\sqrt{2}$, the formal solution of u_0 is written as

$$u_0(y) = A \exp\left[-(1+i)\sqrt{\frac{\omega}{2\nu}} y\right] + B \exp\left[(1+i)\sqrt{\frac{\omega}{2\nu}} y\right]. \tag{5.33}$$

Noting $u_0(y \to \infty) = 0$ we get $B = 0$. Also, in order to satisfy the boundary condition given by Eq. (5.30) we need $A = U_0$. Therefore, the horizontal component of the velocity field (u) is expressed as

$$u(x, y, t) = U_0 \exp\left[-\sqrt{\frac{\omega}{2\nu}} y\right] \cos\left(\omega t - \sqrt{\frac{\omega}{2\nu}} y\right). \tag{5.34}$$

Exercise 5.2 Describe the steady flow in the tube through an equilateral triangular cross-section.

Exercise 5.3 Long cylinder of radius R_1 is rotating with angular speed Ω_1 in an infinite viscous fluid. Assuming perfect symmetry (i.e. $\mathbf{u} = u_\phi(r)\hat{\phi}$), find the expression for the azimuthal velocity component (u_ϕ) of the fluid.

Low Reynolds Number Flows

6

> *[To] mechanical progress there is apparently no end: for as in the past so in the future, each step in any direction will remove limits and bring in past barriers which have till then blocked the way in other directions, and so what for the time may appear to be a visible or practical limit will turn out to be but a bend in the road.*
>
> —*Osborne Reynolds*

In this chapter, we discuss an important regime of viscous flow called "Low Reynolds number flows". We begin by first introducing this very important non-dimensional number in Fluid mechanics. The governing equations for such flows are then derived and their general properties are discussed. Finally, we briefly consider two important applications of low Reynolds number flows in locomotion of micro-organisms and lubrication theory.

6.1 Reynolds Number and Its Significance

The dynamics of incompressible flow characterised by velocity u, density ρ, pressure p and kinematic viscosity ν is governed by Navier-Stokes equation,

$$\frac{\partial u}{\partial t} + u \cdot \nabla u = -\frac{1}{\rho}\nabla p + \nu \nabla^2 u. \tag{6.1}$$

© The Author(s), under exclusive license to Springer Nature Switzerland AG 2022
S. R. Jain et al., *A Primer on Fluid Mechanics with Applications*,
https://doi.org/10.1007/978-3-031-20487-6_6

Assuming typical length-scale and velocity-scale associated with such a flow as L, U respectively, we can estimate corresponding time-scale as $t_0 = L/U$. The non-dimensional quantities (represented by tilde) can now be written as

$$\tilde{u} = \frac{u}{U}, \quad \tilde{t} = \frac{t}{t_0} = \frac{Ut}{L}, \quad \tilde{\nabla} \equiv L\nabla,$$

where $\tilde{\nabla}$ is the dimensionless gradient operator. The Eq. (6.1) can now be written as

$$\frac{U^2}{L}\left[\frac{\partial \tilde{u}}{\partial \tilde{t}} + \tilde{u} \cdot \tilde{\nabla}\tilde{u}\right] = -\frac{1}{\rho L}\tilde{\nabla}p + \frac{\nu U}{L^2}\tilde{\nabla}^2\tilde{u}.$$

Dividing above equation by U^2/L, we get the dimensionless form of the Navier-Stokes equation as

$$\frac{\partial \tilde{u}}{\partial \tilde{t}} + \tilde{u} \cdot \tilde{\nabla}\tilde{u} = -\tilde{\nabla}\tilde{p} + \frac{1}{Re}\tilde{\nabla}^2\tilde{u}. \tag{6.2}$$

Here, dimensionless pressure defined as

$$\tilde{p} = \frac{p}{\rho U^2} \tag{6.3}$$

whereas Re is the Reynolds number given by

$$Re = \frac{UL}{\nu} = \frac{\rho UL}{\mu} \tag{6.4}$$

Reynolds number is considered as one of the most important dimensionless number in the Fluid mechanics. This number guides us in estimating relative importance of inertial term against viscous terms in the Navier-Stokes equation for a given flow. More precisely, Reynolds number is also defined as the ratio of inertial term to viscous term, i.e.,

$$\text{Re} = \frac{\text{Inertial Force}}{\text{Viscous force}} = \frac{\left|u.\nabla u\right|}{\left|\nu\nabla^2 u\right|} = \frac{U^2/L}{\nu U/L^2} = \frac{UL}{\nu} \tag{6.5}$$

It can also be interpreted as the ratio of diffusive time-scale, set-up by the viscosity, to that of advective time-scale of flow. Noting that kinematic viscosity ν has the dimensions L^2T^{-1}, we can estimate the characteristic time-scale of diffusion as L^2/ν. Thus,

$$Re = \frac{\text{Time-scale of diffusion}}{\text{Time-scale of advection}} = \frac{L^2/\nu}{L/U} = \frac{UL}{\nu} \tag{6.6}$$

Thus, in a high Reynolds number ($Re \gg 1$) flow the inertial forces dominate over the viscous force. In such a flow, if we inject a passive scalar (such as dye, ink etc) at a given location in the flow field typically the scalar is transported away from the point of inject by the flow before it gets mixed with the surrounding fluid through the process of diffusion. The situation of course gets more complicated when the Reynolds number becomes very large and the flow field becomes turbulent. Our everyday experience tells us that in a highly turbulent fluid mixing of passive scalar also becomes more efficient. This is typically attributed to enhanced turbulent diffusivity arising out of non-linearity in the inertial forces.

On the contrary, the low Reynolds number ($Re \ll 1$) flows are almost always laminar on account of absence of nonlinear advection term in the governing equation. Interestingly, the typical pressure scale also varies qualitatively with the change in the Reynolds number of the flow. In a high Reynolds number flow, the pressure scales as $p \sim \rho U^2$ as can be seen from (6.3) whereas in flows with low Reynolds number the viscosity appears in the pressure scaling as $p \sim \rho U^2 / Re \simeq \rho U \nu / L \simeq U \mu / L$ as we will see soon. This demonstrates that flow characteristic changes drastically as the Reynolds number changes from low to high.

The Reynolds number plays a significant role in the physics of motion of living species. The Reynolds number associated with motion (swimming/flying) of various species can vary significantly depending upon its size and the characteristic speed of the motion. This can be see from the Table 6.1. From this table, it is clear that the Stokes equations, discussed in the previous section, are valid to described the motion of micro-organism inside the fluid due to extremely small Reynolds number on account of their small size.

In order to highlight the important difference between locomotion at low and high Reynolds number, we consider a swimmer of density ρ_s and characteristic size L moving with speed U inside a fluid of density ρ. Then, the mass of the swimmer scales as $m \sim \rho_s L^3$. Here, a swimmer can be defined as an organism which periodically deforms its body/shape to move forward inside the fluid. We now estimate a typical stopping distance for this swimmer would move after if it stops deforming its shape. In high Reynolds number regime, since pressure scales as ρU^2 the drag force experienced by the swimmer is given by $F_D \sim \rho U^2 L^2$. This amounts to a deceleration of $F_D / m \sim (\rho / \rho_s) U^2 / L$ leading to a stopping distance of $(\rho_s / \rho) L$. For humans, the ratio of $\rho_s / \rho \sim 1$ leading stopping distance of the order

Table 6.1 Typical Reynolds numbers of various species

Species	Size m	Speed (m/(s)	Kinetic viscosity (m^2/s $\times 10^{-6}$)	Reynolds number
Fish (Salmon)	0.75	0.5	1.0	4×10^5
Birds (Eagle)	0.5	20	15	6×10^5
Human (swimming)	2	1	1.0	2×10^6
Micro-organism (E. Coli)	1×10^{-5}	3×10^{-5}	1.0	3×10^{-4}

of size of the swimmer i.e. couple of meters. Interestingly, birds can glide much longer distance compared to their size without flapping the wings due to their higher ρ_s/ρ ratio. On the other hand, in the low Reynolds number regime, the pressure scales as $\rho U v/L$. Therefore, the deceleration experienced goes as $(\rho/\rho_s)U v/L^2$ for which the stopping distance turns out to be $Re(\rho/\rho_s)L$. Since $Re \ll 1$, the stopping distance remains much smaller than the size of the species. Hence, the micro-organism have to continuously deform their shapes in order to move forward in the low Reynolds number regime.

6.2   Stokes Flow (Re ≪ 1)

The flows with low Reynolds numbers (Re≪ 1) are known as Stokes flow or creeping flow. Study of such flows offers unique opportunity to investigate the physics of fluids analytically, mainly due to the absence of nonlinear inertial term in the Navier-Stokes equation. They are also important for number of practical applications such as lubrication theory, sedimention of micro-particles in a highly viscous fluid, locomotion of micro-organisms.

The dimensionless Navier-Stokes equation for an imcompressible fluid, given in (6.2), can be rewritten as

$$\frac{\partial \tilde{u}}{\partial \tilde{t}} + Re\left(\tilde{u} \cdot \tilde{\nabla}\tilde{u}\right) = -\tilde{\nabla}\tilde{p} + \tilde{\nabla}^2\tilde{u}. \tag{6.7}$$

Note that the dimensionless time and pressure are redefined here as

$$\tilde{t} = \frac{1}{Re}\left(\frac{t}{L/U}\right) = \frac{t}{L^2/\nu} \tag{6.8}$$

$$\tilde{p} = Re\left(\frac{p}{\rho U^2}\right) = \frac{p}{\mu U/L} \tag{6.9}$$

Thus, characteristic time-scale for stokes flow is diffusive time-scale whereas the characteristic pressure-scale is set-up by the fluid viscosity and linearly proportional to the flow velocity, unlike the dynamic pressure (ρU^2). Since the pressure is force per unit area, the typical force experienced by a symmetric body of size L can be estimated as $F \sim (\mu U/L)\,L^2 \propto U$. Thus, in low Reynolds number flow, the frictional forces (drag force) becomes proportional to velocity.

Now, taking limiting case of $Re \to 0$ in Eq. (6.7) and going back to dimensional form, we can write the governing equations for the Stokes flow under the influence of external applied force **f** take the form

$$\rho\frac{\partial u}{\partial t} = -\nabla p + \mu\nabla^2 u + \mathbf{f}, \tag{6.10}$$

along with the incompressibility condition of

$$\nabla \cdot \boldsymbol{u} = 0. \tag{6.11}$$

For a conservative applied force $\mathbf{f} = -\nabla\Phi$, by taking $\nabla \times \nabla$ of Eq. (6.10) we can write

$$\rho \frac{\partial}{\partial t} \left(\nabla^2 \boldsymbol{u} \right) = \mu \nabla^4 \boldsymbol{u}, \tag{6.12}$$

where pressure term is eliminated and incompressibility condition is used.

In the absence of applied force, in steady-state these equations reduce to

$$\nabla p = \mu \nabla^2 \boldsymbol{u} = -\mu \nabla \times \boldsymbol{\omega} \tag{6.13}$$

where $\boldsymbol{\omega} = \nabla \times \boldsymbol{u}$ is the vorticity. Note that here we have used the vector identity $\nabla \times \nabla \times \boldsymbol{u} = \nabla (\nabla \cdot \boldsymbol{u}) - \nabla^2 \boldsymbol{u}$ and the incompressibility condition (Eq. (6.11)). Taking curl of above equation and using $\nabla \times \nabla p = \nabla \cdot \boldsymbol{\omega} = 0$, we get the governing equation for the steady-state Stokes flow as

$$\nabla^2 \boldsymbol{\omega} = 0. \tag{6.14}$$

Exercise 6.1 Show that the streamfunction (ψ) in Stokes flow satisfy the biharmonic equation i.e.

$$\nabla^4 \psi = 0 \tag{6.15}$$

6.2.1 Axisymetric Stokes Flow

Let us consider a axisymmetric Stokes flow in the spherical co-ordinate system (r, θ, ϕ). The velocity for such a flow can be represented in terms of stream function (ψ) as

$$\boldsymbol{u} = \nabla\phi \times \nabla\psi \tag{6.16}$$

where $\nabla\phi = \hat{\phi}/r \sin\theta$ with $\hat{\phi}$ as the unit vector in the azimuthal direction (ϕ). The radial and latitudinal components of velocity are then written as

$$u_r = \frac{-1}{r^2 \sin\theta} \frac{\partial \psi}{\partial \theta}, \quad u_\theta = \frac{1}{r \sin\theta} \frac{\partial \psi}{\partial r}. \tag{6.17}$$

The corresponding vorticity is directed in the azimuthal direction, i.e., $\boldsymbol{\omega} = \omega_\phi \hat{\phi}$ which can be written as

$$\boldsymbol{\omega} = \left[\frac{1}{r} \frac{\partial}{\partial r} (r u_\theta) - \frac{1}{r} \frac{\partial u_r}{\partial \theta} \right] \hat{\phi} = \frac{L(\psi)}{r \sin \theta} \hat{\phi} = L(\psi) \nabla \phi \qquad (6.18)$$

where the differential operator L is defined as

$$L \equiv \frac{\partial^2}{\partial r^2} + \frac{\sin \theta}{r^2} \frac{\partial}{\partial \theta} \left(\frac{1}{\sin \theta} \frac{\partial}{\partial \theta} \right) \qquad (6.19)$$

Taking curl of (6.18) we get $\nabla \times \boldsymbol{\omega} = -\nabla \phi \times \nabla [L(\psi)]$. Note that when $\boldsymbol{u} = \nabla \phi \times \nabla \psi$, we got $\boldsymbol{\omega} = \nabla \times \boldsymbol{u} = L(\psi) \nabla \phi$. Similarly,

$$\begin{aligned}
\nabla^2 \boldsymbol{\omega} &= -\nabla \times \nabla \times \boldsymbol{\omega} \\
&= \nabla \times (\nabla \phi \times \nabla L(\psi)) \\
&= L[L(\psi)] \nabla \phi \\
&= L^2 (\psi) \nabla \phi.
\end{aligned}$$

Thus, the governing equation to axisymmetric Stokes flow reduces to differential equation for stream-function ψ, subject to appropriate boundary conditions, in the form

$$L^2 (\psi) = 0 \qquad (6.20)$$

where the differential operator L is defined as per (6.19).

6.2.2   Stokes Drag on a Sphere

We now apply governing equation for axisymmetric Stokes flow, given by (6.20), to compute a drag force on a sphere of radius a placed inside the viscous fluid moving with a vertical velocity $\boldsymbol{U} = -U_o \hat{z}$. Equivalently, we can consider a sphere moving with velocity $\boldsymbol{U} = U_o \hat{z}$ and analyze the motion in the reference frame of moving sphere. The fluid in the vicinity of the sphere gets disturbed causing the change in momentum of fluid as it passes by the sphere. In other words, the sphere exerts a force on the fluid leading to its change in momentum. The equal and opposite reaction force is exerted on the sphere by the fluid. The component of this force in the direction of flow is called as the Drag force.

Let us assume that the size of the sphere and typical velocity of the fluid are such that Re≪ 1 leading to validity of axisymmetric Stokes equations. We apply no-slip boundary condition on the sphere meaning the fluid in contact with the sphere has same velocity as the sphere i.e. $u_r(a, \theta) = u_\theta(a, \theta) = 0$. Thus, at the radius $r = a$, we have

$$\left.\frac{\partial \psi}{\partial \theta}\right|_{(a,\theta)} = \left.\frac{\partial \psi}{\partial r}\right|_{(a,\theta)} = 0. \tag{6.21}$$

On the other hand, at the distance far away from the sphere, the fluid remains undisturbed moving with $U = -U_o\hat{z}$. In spherical co-ordinates with translates into $u_r = -U_0 \cos\theta$ and $u_\theta = U_0 \sin\theta$ in the undisturbed region of the fluid. Therefore, we expect

$$\psi(r, \theta) = \frac{U_0 r^2}{2} \sin^2\theta \quad \text{as} \quad r \to \infty \tag{6.22}$$

to satisfy both u_r, u_θ velocity components in the region far away from the sphere. The the solution of ψ as $r \to \infty$ given by Eq. (6.22) allows seek the general solution of the form

$$\psi(r, \theta) = f(r) \sin^2\theta \tag{6.23}$$

for the Eq. (6.20) which leads to following ordinary differential equation in r:

$$\left[\frac{d^2}{dr^2} - \frac{2}{r^2}\right]^2 f(r) = 0 \tag{6.24}$$

The boundary conditions for this equation can be states as

$$f(a) = 0, \tag{6.25}$$

$$\left.\frac{df}{dr}\right|_{r=a} = 0, \tag{6.26}$$

$$f(r \to \infty) = \frac{U_0 r^2}{2}. \tag{6.27}$$

Note that the first two boundary conditions correspond to $u_r = 0$ and $u_\theta = 0$ respectively whereas the third boundary is obtained from (6.22). Substituting $f(r) = \alpha r^n$ in (6.24) and solving for n, we can write the solution of $f(r)$ in the form

$$f(r) = \frac{A}{r} + Br + Cr^2 + Dr^4. \tag{6.28}$$

Applying boundary conditions given by (6.25) we get the coefficients as

$$A = \frac{U_0 a^3}{4}, \quad B = \frac{-3U_0 a}{4}, \quad C = \frac{U_0}{2}, \quad D = 0.$$

Thus, substituting this solution in (6.23) we get the solution of streamfunction ψ for the uniform flow past sphere in a viscous fluid as

$$\psi (r, \theta) = U_0 r^2 \sin^2 \theta \left(\frac{1}{2} - \frac{3a}{4r} + \frac{a^3}{4r^3} \right) \tag{6.29}$$

The corresponding velocity and vorticity can be written as

$$u_r = -U_0 \left[\frac{a^3}{2r^3} - \frac{3a}{2r} + 1 \right] \cos \theta \tag{6.30}$$

$$u_\theta = -U_0 \left[\frac{a^3}{4r^3} + \frac{3a}{4r} - 1 \right] \sin \theta \tag{6.31}$$

$$\omega = \frac{3U_0 a \sin \theta}{2r^2} \hat{\phi} \tag{6.32}$$

The pressure inside the fluid can be computed by integrating Eq. (6.13) to give

$$p (r, \theta) = p_\infty + \frac{3\mu U_0 a \cos \theta}{2r^2} \tag{6.33}$$

The pressure field $(p - p_\infty)$ for flow past sphere is in low Reynolds number regime is shown with filled-contour plot in Fig. 6.1 where pressure is normalized by $\mu U / a$. The black contour lines represent streamlines for this flow. Thus, one can see that the upstream pressure is higher than the downstream pressure resulting in the net downward drag on the sphere.

Since both the pressure and the velocity field inside the fluid are known, we can now compute the drag force experienced by the sphere. Consider a differential surface element of the sphere given by $ds = a^2 \sin \theta d\theta d\phi$ pointing radially out. The components of the stresses experienced by the fluid in contact with this differential surface element are given by

$$\sigma_{rr} (a, \theta) = \left(-p + 2\mu \frac{\partial u_r}{\partial r} \right)_{r=a} = -p_\infty - \frac{3\mu U_0}{2a} \cos \theta \tag{6.34}$$

$$\sigma_{r,\theta} (a, \theta) = \mu \left(\frac{1}{r} \frac{\partial u_r}{\partial \theta} + \frac{\partial u_\theta}{\partial r} - \frac{u_\theta}{r} \right)_{r=a} = \frac{3}{2} \frac{\mu U_0}{a} \sin \theta \tag{6.35}$$

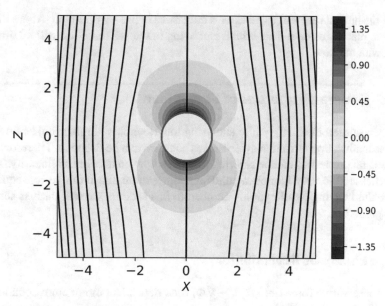

Fig. 6.1 Stokes flow passing over a sphere. The pressure field is represented by continuous contours whereas streamlines are shown with lines

where we have used velocity and pressure equations given by Eqs. (6.30), (6.31), and (6.33). Therefore, noting $f_i = \sigma_{ij} n_j ds$ and substituting shear stresses from above, we can compute the force experienced by this differential element as

$$d\boldsymbol{f} = a^2 \sin\theta d\theta d\phi \left[\left(-p_\infty - \frac{3\mu U_0}{2a} \cos\theta \right) \hat{\boldsymbol{r}} + \left(\frac{3}{2} \frac{\mu U_0}{a} \sin\theta \right) \hat{\boldsymbol{\theta}} \right]$$

Now, the drag force on the element can be calculated by $df_z = d\boldsymbol{f} \cdot \hat{\boldsymbol{z}}$ which gives

$$df_z = a^2 \sin\theta d\theta d\phi \left[-p_\infty \cos\theta - \frac{3\mu U_0}{2a} \right].$$

Finally, total drag force on the sphere is given by

$$F_z = \int_0^{2\pi} d\phi \int_0^\pi \left(-a^2 p_\infty \cos\theta - \frac{3}{2} \mu U_0 a \right) \sin\theta d\theta.$$

Solving this integral we get the famous formula for Stokes drag as

$$F_z = -6\pi \mu U_0 a. \tag{6.36}$$

Note that the negative sign in the above equation is consistent with the choice of our fluid input flow velocity. This implies that the sphere moving in the $\hat{z}$ direction inside the fluid at rest will experience a drag force in the $-\hat{z}$ direction. Alternatively, a sphere placed inside a fluid which is moving in the $-\hat{z}$ direction will be dragged along with the flow.

6.3   General Properties of Stokes Flow

As we have seen in (6.10), while analysing low Reynolds number Stokes flow; the nonlinear advection term in Navier-Stokes equation can be dropped. Therefore, certain general properties can be asserted to Stokes flow on account of linearity. These properties lead to very interesting and sometimes quite non-intuitive consequences for the fluid having low Reynolds number. In this section, we will discuss some of those properties.

6.3.1   Kinematic Reversibility

For a conservative force field, $\mathbf{f} = -\nabla \Phi$, time-dependent momentum equation for Stokes flow can be written as

$$\rho \frac{\partial \boldsymbol{u}}{\partial t} = -\nabla p + \mu \nabla^2 \boldsymbol{u} - \nabla \Phi, \qquad (6.37)$$

Taking divergence of above equation we get an equation for the pressure in the form

$$\nabla^2 p = -\nabla^2 \Phi. \qquad (6.38)$$

Thus, on reversal of force field ($\Phi \rightarrow -\Phi$), both pressure ($p \rightarrow -p$) and velocity field get reversed ($\boldsymbol{u} \rightarrow -\boldsymbol{u}$). Such time-reversibility associated with Stokes flow is an important consequence of the linearity of Stokes equations. In other words, in a highly viscous fluid the velocity field traces back its history on reversal of applied force, $\mathbf{f}$. For example, a drop of dye which is well-mixed inside a highly viscous flow through controlled stirring will be completely unmixed on exact time reversal of stirring action on the fluid.

Time reversal along with mirror symmetry can be employed to draw some interesting conclusions about the motion of objects inside the low Reynolds number flow without performing any actual calculation. In order to demonstrate this point, we consider a simple model for microorganism having spherical head and a rigid tail as shown in Fig. 6.2. This species aims propel inside the viscous fluid having $Re \ll 1$. Using kinetic time-reversibility and mirror symmetry, we can show that the micro-organism can't swim inside the fluid by flapping its tail rigidly up and down. For this, we first assume that the species is instantaneously moving towards left with a speed U by moving its tail downwards as shown in Fig. 6.2a. Now, applying mirror symmetry, we find that species will move towards right with the same stroke (i.e. downwards) in the mirror world represented by Fig. 6.2b. Now, we

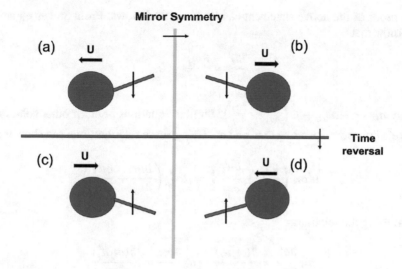

Fig. 6.2 Application of mirror symmetry (left-right) and time reversal (top-bottom) to demonstrate that a micro-organism having spherical head and rigid tail can not perform locomotion inside a low Re fluid

apply time-reversibility to get corresponding figures Fig. 6.2c, d respectively. Now, comparing Fig. 6.2b, c and a–d we find that U must be zero since strokes of the tail in different directions is causing propulsion in the same direction which clearly is a contradiction. Thus, with such simple arguments employing time reversibility one can rule out propulsion of micro-organism by simple periodic motion of rigid tail with one degree of freedom.

Exercise 6.2   Use arguments of mirror symmetry and time reversal to show that a sedimenting particle travelling parallel to a wall experiences zero force normal to the wall.

6.3.2   Reciprocal Theorem

Another important consequence of linearity having wide applicability in study of Stokes flow is the reciprocal theorem. This is theorem allows us to infer properties of some unknown Stokes flow solution from known solution of Stokes flow in the same domain. Consider two Stokes flow solutions inside a region V bounded by a closed surface S. In this region, let $\mathbf{u}$ and $\mathbf{u}'$ represent two velocity fields associated with stress fields $\boldsymbol{\sigma}$ and $\boldsymbol{\sigma}'$ respectively. Then according to reciprocal theorem

$$\int_{S} \mathbf{u} \cdot (\boldsymbol{\sigma}' \cdot \hat{\mathbf{n}}) \, ds = \int_{S} \mathbf{u}' \cdot (\boldsymbol{\sigma} \cdot \hat{\mathbf{n}}) \, ds. \qquad (6.39)$$

The proof of the above statement can be given as follows: From Stokes equations we know that

$$\frac{\partial u_i}{\partial x_i} = \frac{\partial \sigma_{ij}}{\partial x_j} = 0,$$

where $\sigma_{ij} = -p\delta_{ij} + \mu \left(\frac{\partial u_i}{\partial x_j} + \frac{\partial u_j}{\partial x_i} \right)$. Similar equations hold for other solution of Stoke's flow corresponding to $\mathbf{u}'$ and σ'. Using these relations, one can show that

$$\mu' \sigma_{ij} \left(\frac{\partial u'_i}{\partial x_j} + \frac{\partial u'_j}{\partial x_i} \right) = \mu \sigma'_{ij} \left(\frac{\partial u_i}{\partial x_j} + \frac{\partial u_j}{\partial x_i} \right)$$

Now, using the relations

$$\sigma_{ij} \frac{\partial u'_i}{\partial x_j} = \frac{\partial (\sigma_{ij} u'_i)}{\partial x_j} - u'_i \frac{\partial \sigma_{ij}}{\partial x_j} = \frac{\partial (\sigma_{ij} u'_i)}{\partial x_j},$$

$$\sigma'_{ij} \frac{\partial u_i}{\partial x_j} = \frac{\partial (\sigma_{ij} u_i)}{\partial x_j} - u_i \frac{\partial \sigma'_{ij}}{\partial x_j} = \frac{\partial (\sigma'_{ij} u_i)}{\partial x_j}$$

we can show that

$$\mu' \frac{\partial (\sigma_{ij} u'_i)}{\partial x_j} = \mu \frac{\partial (\sigma'_{ij} u_i)}{\partial x_j}.$$

Integrating above equations over volume V and applying divergence theorem we prove the reciprocal theorem for $\mu = \mu'$ as

$$\int_S (\sigma' \cdot \mathbf{u}) \, d\mathbf{s} = \int_S (\sigma \cdot \mathbf{u}') \, d\mathbf{s}.$$

which is equivalent to (6.39).

6.3.3 Force and Torque on a Body of Arbitrary Shape Inside a Low Re Flow

The linear relationship between force and the velocity for the Stokes flow is a consequence of linearity associated with the Stokes equations (6.10) and (6.11). For a body of arbitrary shape, the force and the velocity are related through a rank two tensor which is independent of velocity. In fact, general expression for the force

F and Torque **T** on the body undergoing translational (**U**) and rotation (Ω) inside the low Re flow can be written as

$$F_i = A_{ij}U_j + B_{ij}\Omega_j \tag{6.40}$$

$$T_i = C_{ij}U_j + D_{ij}\Omega_j \tag{6.41}$$

Using reciprocal theorem, we can show that $A_{ij} = A_{ji}$, $D_{ij} = D_{ji}$ and $B_{ij} = C_{ji}$. Therefore above relations can be expressed in the matrix form as [55]

$$\begin{pmatrix} \mathbf{F} \\ \mathbf{T} \end{pmatrix} = \begin{pmatrix} \mathbf{A} & \mathbf{B} \\ \mathbf{B^T} & \mathbf{D} \end{pmatrix} \cdot \begin{pmatrix} \mathbf{U} \\ \mathbf{\Omega} \end{pmatrix}. \tag{6.42}$$

Here, **A**, **B**, **D** are called resistance matrices. The above relations can be inverted to express **U**, Ω in terms of **F**, **T** through corresponding mobility matrices. For a symmetric bodies, **A**, **D** are diagonal matrices whereas matrix **B** vanishes. This implies that a symmetric body inside such a fluid can not experience a torque while performing pure translational motion. Equivalently, for a symmetric body undergoing pure rotation the linear drag force has to go to zero. For a general asymmetric body, the resistance matrices **A**, **D** can have off-diagonal components which can lead to anisotropic drag force or torque. Thus, a body of general shape can generate a differential drag force by arbitrarily deforming itself to generate the required flow field. This property plays a crucial role in the locomotion of micro-organisms inside low Reynolds number fluid. Another important consequence of (6.42) is occurrence of translational motion through angular forcing or vice versa when the matrix **B** is non-zero. This plays a crucial role in locomotion of chiral bodies, for example motion of a bacteria through rotating flagella.

6.4   Scallop Theorem for Locomotion in Micro-Organisms

We have seen that the Reynolds number associated with micro-organisms swimming in a water-like fluid can be as low as $10^{-5} - 10^{-6}$. Therefore, Stokes equations are valid to describe the locomotion of micro-organisms. One of the most celebrated result in Stokesian locomotion is known as *Scallop theorem* which by first proposed by Purcell [94]. According to the theorem, a Stokesian swimmer undergoing periodic deformation which is identical under time reversal can not generate a net thrust for achieving propulsion. This is a direct consequence of kinematic reversibility of Stokes flow. In the previous section, we have seen that the property of kinematic reversibility implies that on reversal of time every fluid parcel exactly traces back its history when Reynolds number is zero. Thus, flow field and associate pressure will reverse in every half-cycle of the periodic deformation leading to zero net (cycle-averaged) force on the micro-organism.

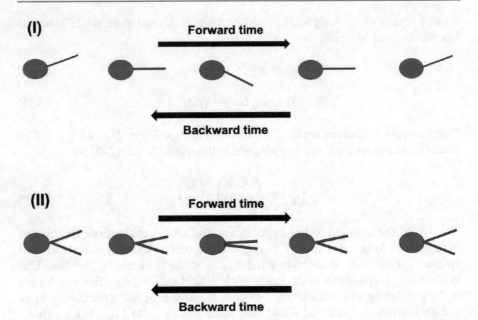

Fig. 6.3 Application of scallop theorem to demonstrate that species with single rigid tail (case I) and two rigid tails (case II) can not perform locomotion inside a low *Re* fluid by undergoing time-reversible flipping of the tail(s). Note that both sequences are identical in forward and backward time direction

Let us consider a periodic motion of micro-organism having rigid head and tail which was described in the previous session. Through the application of time reversibility and mirror symmetry, we had shown that this prototypical micro-organism (see Fig. 6.2) can't swim by transverse flapping of the tail. We can draw identical conclusion from scallop theorem as shown in case I of Fig. 6.3. We see that the scallop theorem will apply here since the motion is identical under the time reversal operation and locomotion is not possible. Even for a micro-organism having two-link tail (scallop-like) described in the case-II of Fig. 6.3 similar conclusions can be drawn. On the contrary, three link tail micro-organism described in Fig. 6.4 can generate thrust for propulsion as this deformation is not time reversible. This can be seen by comparing position (b) and (d) of the cycle. Note that although individual links in this case are still performing time reversible motion but due to phase difference between various links overall time reversibility is lost. Therefore, this species can experience a cycle-averaged thrust force.

6.4.1 Flagellar Locomotion

Taking clue from three-link tail species described in Fig. 6.4, we consider a prototypical organism having a rigid spherical head but flexible tail, typically known

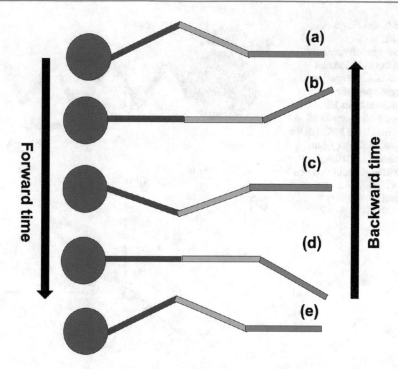

Fig. 6.4 Three link-tail species can perform locomotion through its time irreversible flipping of the tail. Note that the sequences are not identical in forward and backward time direction

as flagella, as shown in the Fig. 6.5a. The species aims to achieve motion towards left by producing planar waves in the fluid (towards right) through a periodic deformation that can be described in the following form

$$y(t) = a_0 \sin (kx - \omega t), \tag{6.43}$$

y denotes transverse displacement of flagella, $k = 2\pi/\lambda$ is the wavenumber and ω is the frequency. Note that the deformation is modelled in the form of propagating wave travelling towards right ($+\hat{x}$ direction) with speed $c = \omega/k$. Let us estimate the steady-state swimming velocity $-U\hat{x}$ due to momentum imparted by fluid to the micro-organism. Consider a differential section ds of the flagella between the planes $A - A'$ and $B - B'$ shown in Fig. 6.5a. The velocity of this element can be written as

$$\mathbf{V} = -U\hat{x} + \frac{\partial y}{\partial t}\hat{y} \tag{6.44}$$

Fig. 6.5 (a) Flagellar locomotion of micro-organism modelled with spherical head and flexible tail. The species undergoes periodic deformation of its tail to produce planar sinusoidal waves inside the fluid. (b) the drag forces along (f_s) and transverse (f_t) to the differential element between $A - A'$ and $B - B'$ are not identical

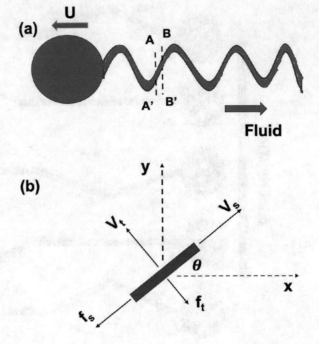

The components of velocity along and normal to this element as represented by V_s and V_t respectively as shown in Fig. 6.5b. These components can be expressed as

$$V_s = -U \cos \theta + \left(\frac{\partial y}{\partial t} \right) \sin \theta,$$

$$V_t = U \sin \theta + \left(\frac{\partial y}{\partial t} \right) \cos \theta$$

where θ is angle made by the differential section with the x−axis. The corresponding drag forces per unit length can be calculated as $\mathbf{f}_s = -\xi_s \mathbf{V}_s$ and $\mathbf{f}_t = -\xi_t \mathbf{V}_t$ where ξ_s, ξ_t are the drag coefficients in the directions parallel and perpendicular to the differential element as shown in Fig. 6.5b. Substituting above relations for V_s, V_t and using small angle approximation, we can write the horizon (f_x) and vertical (f_y) component of drag force on the element in the following form:

$$f_x = \xi_s U + (\xi_t - \xi_s) \left(\frac{\partial y}{\partial t} \right) \left(\frac{\partial y}{\partial x} \right), \tag{6.45}$$

$$f_y = -\xi_t \left(\frac{\partial y}{\partial t} \right) + (\xi_s - \xi_t) \left(\frac{\partial y}{\partial x} \right) U. \tag{6.46}$$

Note that we have used $\sin\theta \simeq \tan\theta = \partial y/\partial x$, consistent with small angle approximation. The net horizontal force F_x on the flagella of length $L = N\lambda$ is given by $F_x = \int_0^{N\lambda} f_x dx$. Now, substituting f_x from (6.45) and using (6.43) we get

$$F_x = \int_0^{N\lambda} \left[\xi_x U - (\xi_t - \xi_s) \left(a_0^2 k^2 c \right) \cos^2 (kx - \omega t) \right] dx.$$

Integrating above equation, we get the total instantaneous horizontal force on flagella as

$$F_x = N\lambda \left[\xi_s U - \left(\frac{a_0^2 k^2 c}{2} \right) (\xi_t - \xi_s) \right], \tag{6.47}$$

where the second term represents the contribution due to anisotropic drag-coefficients discussed in the previous section. Since the micro-organism is moving with a constant speed U along $-\hat{x}$ direction, this force should be balanced by the drag force (F_H) on its head, i.e., $F_x = -F_H$. Noting for a spherical head with radius R_0, the drag force can be written as $F_H = \xi_H U R_0$ where ξ_H is the drag-coefficient which can be computed from Stokes drag formula derived earlier. The swimming speed U can now be computed from this force-free condition as

$$U = \left(\frac{a_0^2 k^2 c}{2} \right) \left(\frac{\xi_t}{\xi_s} - 1 \right) \left(\frac{1}{1 + \frac{R_0 \xi_H}{N\lambda \xi_s}} \right) \tag{6.48}$$

Note that for isotropic drag-coefficient i.e. $\xi_t = \xi_s$, the swimming speed vanishes which emphasizes the important role of drag-anisotropy in the locomotion in low Reynolds number fluid.

6.5 Low Reynolds Number Flows in Lubrication

Another example for applicability of low Reynolds number regime can be found in the phenomenon which takes place in oil lubricated bearings having a viscous flim of thickness $h \ll l$, affords an example in which the effect of viscosity predominates over the inertial forces in the equation of motion,

$$\rho \mathbf{u} \cdot \nabla \mathbf{u} = -\nabla p + \mu \nabla^2 \mathbf{u}, \tag{6.49}$$

μ being the coefficient of viscosity. For a thin layer ($h \ll l$), the largest viscous term in the equation of motion for the x-direction is $\mu \frac{\partial^2 u}{\partial z^2}$. Here l is the length-scale of flow variation in the xy-direction and h is the typical thickness of the viscous flow.

Thus, for $h \ll l$, the ratio of inertial to viscous force,

$$\frac{\rho \left(u \partial u / \partial x \right)}{\mu \left(\partial^2 u / \partial z^2 \right)} = \frac{\rho U^2 / l}{\mu U / h^2} = \frac{\rho U l}{\mu} \left(\frac{h}{l} \right)^2 = R \left(\frac{h}{l} \right)^2 \ll 1. \tag{6.50}$$

$R = \rho U l / \mu = U l / \nu$ is the Reynolds number, with ν being the kinematic viscosity. Let us now derive the Reynolds equation for motion in a thin viscous film for small Reynolds number, $R \ll 1$. Neglecting the inertial terms, the equation of motion for an incompressible fluid layer in a steady-state becomes $\nabla p = \mu \nabla^2 \mathbf{u}$, which implies $\nabla^2 p = \mu \nabla^2 \left(\nabla \cdot \mathbf{u} \right) = 0$, for an incompressible fluid.

For a slow viscous motion, with $R \ll 1$, this reduces to $\nabla p = \mu \frac{\partial^2 u}{\partial z^2}$, neglecting the inertial terms. Let U be the scale of motion in the xy-plane and W that in the z-direction. Then, from the continuity equation for an incompressible fluid, $\nabla \cdot \mathbf{u} = 0$, we recover, $U / l \simeq W / h$ or $W \simeq \frac{hU}{l} \ll U$, and in fact for $h / l \ll 1$, we have $W \ll U h / l \ll 1$.

Note the x-component of the momentum equation gives

$$\rho \mathbf{u} \cdot \nabla u = \frac{\partial p}{\partial x} + \mu \frac{\partial^2 u}{\partial z^2} \tag{6.51}$$

and hence the inertial term $\equiv O \left(\rho U^2 / l \right) = O \left(\rho W U / h \right)$ and the viscous term $\equiv O \left(\mu U / h^2 \right)$. Clearly, $\rho U^2 / l \ll \mu \rho U / h^2$, provided $U h^2 / l \nu \ll 1$. Now, $U / l \simeq W / h$ from the continuity equation.

Thus, $U h^2 / l \nu \simeq W h / \nu$, and

$$\frac{Wh}{\nu} = \left(\frac{Ul}{\nu} \right) \left(\frac{h}{l} \right) \left(\frac{W}{U} \right) \ll 1,$$

since $W \ll U, R \ll 1, h \ll l$. The momentum equation in the x-direction then becomes

$$\frac{\partial p}{\partial x} = \mu \frac{\partial^2 u}{\partial z^2}, \tag{6.52}$$

Likewise, the momentum equation in the y-direction becomes

$$\frac{\partial p}{\partial y} = \mu \frac{\partial^2 v}{\partial z^2}, \tag{6.53}$$

Let us now consider the momentum equation in the z-direction,

$$\rho \mathbf{u} \cdot \nabla w = -\frac{\partial p}{\partial z} + \mu \frac{\partial^2 w}{\partial z^2} \tag{6.54}$$

The inertial terms on the LHS, $O\left(\rho u \frac{\partial w}{\partial x}\right)$ and $O\left(\rho w \frac{\partial w}{\partial z}\right) = O\left(\rho W^2/h\right)$ and $O\left(\rho U W/l\right)$ and the viscous term $= O\left(\mu W/h^2\right) \ll O\left(\mu U/h^2\right)$, since $W \ll U$. Further note that $\rho W/h \ll \mu W/h^2$, since $Wh\nu \ll 1$. The momentum equation in the z-direction finally yields

$$\frac{\partial p}{\partial z} = 0, \text{ or } p \equiv p\,(x, y) \tag{6.55}$$

and $\partial p/\partial x$ and $\partial p/\partial y$ are independent of z. The forgoing equations can be integrated to give

$$u = \frac{1}{2\mu}\frac{\partial p}{\partial x}z\,(z - h) + \frac{Uz}{h} \tag{6.56}$$

$$v = \frac{1}{2\mu}\frac{\partial p}{\partial y}z\,(z - h) + \frac{Vz}{h}, \tag{6.57}$$

satisfying the boundary conditions $u = 0, v = 0$ at $z = 0$ and $u = U, v = V$ at $z = h$.

The flux out of the column of unit cross-section $dxdy$ and height, h then is given by

$$= \frac{\partial}{\partial x}\int_0^h u\,dz + \frac{\partial}{\partial y}\int_0^h v\,dz$$

$$= \text{rate of decrease of volume of the column}$$

$$= -\frac{\partial h}{\partial t}$$

Now,

$$\int_0^h u\,dz = -\frac{1}{12\mu}\frac{\partial p}{\partial x}h^3 + \frac{1}{2}Uh,$$

$$\int_0^h v\,dz = -\frac{1}{12\mu}\frac{\partial p}{\partial y}h^3 + \frac{1}{2}Vh$$

The equation of the bounding surface is, $z = h\,(x, y)$ and at the upper surface, $\frac{d}{dt}\,(h - z) = 0$, i.e.

$$\frac{\partial h}{\partial t} + U\frac{\partial h}{\partial x} + V\frac{\partial h}{\partial z} - W = 0,$$

But,

$$\frac{\partial h}{\partial t} = -\frac{1}{12\mu}\left[\frac{\partial}{\partial x}\left(h^3\frac{\partial p}{\partial x}\right) + \frac{\partial}{\partial y}\left(h^3\frac{\partial p}{\partial y}\right)\right] + \frac{1}{2}\left[\frac{\partial}{\partial x}\left(Uh\right) + \frac{\partial}{\partial y}\left(Vh\right)\right]$$

which gives

$$\frac{\partial}{\partial x}\left(h^3\frac{\partial p}{\partial x}\right) + \frac{\partial}{\partial y}\left(h^3\frac{\partial p}{\partial y}\right) = 6\mu\left[h\left(\frac{\partial U}{\partial x} + \frac{\partial V}{\partial y}\right)\right] + 6\mu\left[U\frac{\partial h}{\partial x} + V\frac{\partial h}{\partial y}\right]$$

$$+ 12\mu W - 12\mu\left[U\frac{\partial h}{\partial x} + V\frac{\partial h}{\partial y}\right]$$

or,

$$\frac{\partial}{\partial x}\left(h^3\frac{\partial p}{\partial x}\right) + \frac{\partial}{\partial y}\left(h^3\frac{\partial p}{\partial y}\right) = 6\mu\left[h\left(\frac{\partial U}{\partial x} + \frac{\partial V}{\partial y}\right) - \left(U\frac{\partial h}{\partial x} + V\frac{\partial h}{\partial y}\right) - 2W\right]$$

$$(6.58)$$

This is the celebrated Reynolds equation for viscous motion in a thin layer of fluid.

Exercise 6.3 With what force does a circular rubber stamp has to be pulled vertically up from the ink pad !!

Physiological Hydrodynamics

7

> *I am not one and simple, but complex and many.*
>
> —*Virginia Woolf*

We discuss the blood flow in our body, a subject which has been treated theoretically and experimentally for a long time [75]. We do not discuss other important topics like air flow in lungs, urinary flow in kidneys, and so on. Before coming to the biofluid mechanics [64, 117], we introduce relevant physiological details. It is important to understand the system in some detail to appreciate how complicated the circulatory system is.

7.1 Basics: Blood Flow Along Arteries

Every minute, human heart pumps approximately five litres of blood. Complete circulation involves exchange of gas between the metabolism in the human tissue and the air in the atmosphere. Systemic circulation begins with aorta which branches into large arteries which, in turn, branch further into capillaries. In capillaries, blood gives up oxygen and takes in carbon dioxide. From here, the blood flows out to the veins. On the other hand, in the pulmonary circulation, blood gives up part of its carbon dioxide and takes in as much oxygen as it had previously given up to the tissues.

The Reynolds number for the blood flow is somewhere between a hundred and a thousand. Beating of heart causes a periodic laminar flow in the smaller arteries, and a transitional flow in larger arteries. The first Fourier component of the velocity distribution of the blood pulse has an angular frequency, ω which critically depends upon the ratio of the arterial radius R and the thickness of the oscillating boundary layer. Taking the viscosity, ν of blood as 4×10^{-6} m^2/s and the angular frequency of the pulse, ω equal to 8 Hz, the thickness of the boundary layer is $\delta \sim \sqrt{\nu/\omega} \sim 0.7$

© The Author(s), under exclusive license to Springer Nature Switzerland AG 2022   99
S. R. Jain et al., *A Primer on Fluid Mechanics with Applications*,
https://doi.org/10.1007/978-3-031-20487-6_7

mm. Thus, boundary layer thickness is much lesser than the radius of the artery. This implies that the velocity distribution across the cross-section is almost uniform.

Pulse in the aorta has an expansion velocity of about 5 m/s. The pressure and velocity pulse is a superposition of the original travelling wave and the reflected waves created at branching of arteries. Thus the wave in the artery is something intermediate between a travelling wave and a standing wave. This means that the aorta acts as a volume reservoir for the output of the heart and ensures an almost steady volume flux of the blood circulation. A brief word about the constitution of blood. Blood plasma has 90% water which has Newtonian properties. However, there is 40–50% of volume suspension of deformable blood corpuscles. The RBC are deformable disk-like bodies approximately 8 microns across. This suspension endows blood with non-Newtonian properties. We will ignore this complication and work with some "effective viscosity" for blood.

Heart has two separate pump chambers–left and right ventricles and atria, made of cardiac muscle. Systemic circulation begins at the right atrium (RA) where the blood is weak in oxygen. From there, it goes to right ventricle (RV) which empties into pulmonary circulation (PC). From here, re-oxygenated blood from PC goes to left atrium (LA), then via left ventricle (LV) to the systemic ciculation. This cycle continues. Atria and ventricles are separated by atrioventricular valves. RV has three flaps—a tricuspid valve whereas LV has two flaps—a mitral valve. Flaps ensure that the atria can fill with blood between heartbeats, they also prevent backflow of blood while ventricles contract. While ventricles relax, the pulmonary valve prevents backflow of blood out of the aorta into the LV. During one cardiac cycle, ventricles carry out a periodic contraction. Cardiac cycle consists of four phases: (A) Isovolumetric ventricle contraction (filling phase), (B) Isovolumetric ventricle contraction (contraction phase), (C) Isovolumetric ventricle relaxation (evacuation phase), (D) Isovolumetric ventricle relaxation (relaxation phase). Steps (A) and (D) define a diastole whereas (B) and (C) define a systole. Ventricle is filled during phase (D). At this point, pressure is only slightly higher in LA than in LV. Therefore, mitral valve is open. This implies that blood flows from pulmonary veins to LA to LV. As the filling volume increases and ventricle expands, pressure in ventricle increases. Pressure in aorta is much larger, so the aortal valve remains closed. The arterial pressure decreases continuously during diastole, corresponding to blood flow in arterial vascular system. With ventricle contraction starting, ventricle pressure increases above atrium, so bicuspid valve closes, implying that the ventricle has to retain a constant blood volume. Ventricle pressure goes to 166 mbar but pressure in arteries decrease continuously. Aortal valve is open when ventricle pressure is greater than aortal pressure. Now a constant quantity of blood is forced into aorta, leading to an increase in aortal pressure from a minimum of 107 mbar to maximum 160 mbar. After ventricle relaxation has begun, the ventricle pressure goes down to less than arterial pressure, implying the closure of aortal and pulmonary valve. The phase of isovolumetric relaxation now follows. The first phase of the diastole goes on as long as the ventricle pressure is lesser than the atrium pressure. Thus the bicuspid valve opens, and next cycle begins.

Systemic circulation can be split into blood circulation system, consisting of aorta, large and small arteries and arterioles. These further divide into capillaries, where gas an material exchange takes place in the microcirculation by diffusion. The blood flows back to the heart via the venules, small and large veins, and the vena cava.

Mean blood pressure on leaving the LV is about 133 mbar. This drops to 13 mbar when blood returns to RV. Due to elastic properties of aorta, pressure is between 120 mbar and 160 mbar around the mean value. In larger arteries, the amplitude of pulsation increases because of wave reflection. Because of elasticity, aorta and large arteries act as volume reservoir. The acceleration part of the blood pulse is reduced, and a higher pressure level is retained during diastole and systole. Thus the flow in arterial branches is smoother. The Reynolds number R of the blood flow in various parts is 3400 (aorta), 500 (large arteries), 0.7 (arterioles), 0.002 (capillaries), 0.01 (venules), 140 (large veins), 3300 (Vena Cava).

Blood has about 90% of plasma which is water along with proteins, antibodies, fibrinogen. Mean volume of blood in about 5 L (man), and, 4 L (woman). Of the total volume, 84% consists of systemic vascular system, 9% for pulmonary circulation, and 7% for heart. Shear rates are governed by the suspension. Viscosity of blood can be considered only if the suspension occurs as a homogeneous liquid. This is true for blood flow in large vessels. In smaller vessels (capillaries), the elastic RBC with their diameter of 8 μm have to be considered as an inhomogeneity. The plot [83] of viscosity against the shear rate, $\dot{\gamma} = \partial u / \partial r$ reveals different kinds of aggregation (Fig. 7.1). This variation has been modelled in terms of a relation between shear

Fig. 7.1 The "low" shear rates correspond to backflows, aggregation occurs leading to unhealthy circulation. The "healthy" part in this plot corresponds to a shear stress of the order ~ 50 N/m^2. RBC begin to pull apart in a spindle-like manner, and no aggregation occurs

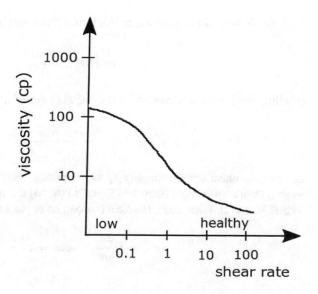

stress, τ and shear rate by the Casson's equation:

$$\sqrt{\tau} = K\sqrt{\dot{\gamma}} + \sqrt{C} \tag{7.1}$$

where K is Casson's viscosity and C denotes the deformation stress of the blood. Fitting to experimental results:

$$\sqrt{\frac{\tau}{\mu_p}} = 1.53\sqrt{\dot{\gamma}} + 2, \tag{7.2}$$

with μ_p the plasma viscosity. For $\dot{\gamma} > 100$, the flow is Newtonian.

7.2 Blood Flow, Pumped by Human Heart

7.2.1 Response of Arterial Walls to Pressure

Consider a cylindrical elastic tube which contains some fluid whose pressure (not necessarily constant) is known [117]. Let us assume that the pressure doesn't vary along the length of the tube. How does the tube respond?

Consider the stresses on an infinitesimal arterial section of length dz. Due to azimuthal symmetry, the radial forces can be written as:

$$F_r = -\left[-\sigma_{rr}(rd\theta dz) + (\sigma_{rr} + d\sigma_{rr}((r+dr)d\theta dz) - 2\sigma_{\theta\theta}\frac{d\theta}{2}drdz \right] \tag{7.3}$$

With the density of the material, ρ, this radial force will be

$$F_r = \rho r dr d\theta dz \frac{d^2 u_r}{dt^2}, \tag{7.4}$$

entailing the equation of motion for the radial component arterial displacement:

$$\rho\frac{d^2 u_r}{dt^2} = -\frac{\partial \sigma_{rr}}{\partial r} + \frac{\sigma_{\theta\theta} - \sigma_{rr}}{r}. \tag{7.5}$$

Let us now make certain simplifying assumptions about the nature of the arterial wall: (i) walls are elastic, obey the Hooke's law, (ii) the material is in compressible so that $\nabla.\mathbf{u} = 0$. With these, Hooke's law can be expressed through

$$\sigma_{rr} = Y\frac{\partial u_r}{\partial r}, \quad \sigma_{\theta\theta} = Y\frac{u_r}{r} \tag{7.6}$$

Eq. (7.5) becomes

$$\rho \frac{d^2 u_r}{dt^2} = -Y \frac{\partial^2 u_r}{\partial r^2} - \frac{Y}{r} \left(\frac{\partial u_r}{\partial r} - \frac{u_r}{r} \right). \tag{7.7}$$

We need to solve for u_r.

Example If internal pressure is constant, then LHS of (7.7) is zero. RHS is the second-order equation with boundary conditions on u_r: $\sigma_{rr} = -P$ for $r =$ inner radius of the tube , R_i, and, zero for $r = R_o$ (outer radius of tube).

The general solution of RHS of (7.7) is

$$u_r = Ar + \frac{B}{r}, \tag{7.8}$$

where A, B are constants. $A = B/R_o^2$ is found by demanding $\sigma_{rr} = 0$ at $r = R_o$. At R_i, $\sigma_{rr} = -P$. This leads to

$$B = \frac{P}{Y} \left(\frac{1}{R_i^2} - \frac{1}{R_o^2} \right)^{-1}. \tag{7.9}$$

The solution is

$$u_r = \frac{P R_i^2 R_o^2}{Y(R_o^2 - R_i^2)} \left(\frac{r}{R_o^2} - \frac{1}{r} \right). \tag{7.10}$$

The stress exerted axially around the tube at its outer boundary is

$$\sigma_{\theta\theta}(r = R_o) = \frac{Y}{R_o} u_r(r = R_o) = \frac{2P}{\left(\frac{R_o^2}{R_i^2} \right) - 1}. \tag{7.11}$$

The blood is pumped by the heart, the periodic pumping creates a time-dependent pressure. Mathematically, nothing much changes. The component, u_r is a function of t also. We may write the time-dependent part as $T(t) = T_0 e^{i\lambda t}$. The radial dependence is embodied in $R(r)$, given from the solution is

$$\frac{d^2 R}{dr^2} + \frac{1}{r} \frac{dR}{dr} - \left(1 - \frac{\lambda^2 \rho r^2}{Y} \right) \frac{R}{r^2} = 0. \tag{7.12}$$

For blood flow, $\lambda \sim 7$ rad/s, blood density $\rho \sim 1.1$ g/cc. For radius, 1 cm and $Y \sim 10^6$ dynes/cm^2, we have $\frac{\lambda^2 \rho r^2}{Y} \sim 10^{-5} \ll 1$. The above equation is thus simplified, the solution is easily seen as

$$u_r = \sum_\lambda a_\lambda \left(Ar + \frac{B}{r} \right) e^{i\lambda t} \tag{7.13}$$

with boundary conditions, $\sigma_{rr} = -P$ at $r = R_i$, $\sigma_{rr} = 0$ at $r = R_0$. Time-dependent pressure can be written as a Fourier expansion,

$$P(t) = \sum_n P_n e^{i\omega_n t}, \qquad \omega_n = \frac{2\pi n}{\tau} \tag{7.14}$$

where τ is the period of the pulse. Imposing the boundary conditions at the inner surface, we obtain

$$B \left(\frac{1}{R_o^2} - \frac{1}{R_i^2} \right) \sum_\lambda a_\lambda e^{i\lambda t} = -\frac{1}{Y} \sum_n P_n e^{i\omega_n t}. \tag{7.15}$$

Since this is valid for all t, this gives a_λ, and, $\lambda = \omega_n$. The final solution is

$$u_r = \frac{R_i^2 R_o^2 P(t)}{Y(R_o^2 - R_i^2)} \left[\frac{r}{R_o^2} + \frac{1}{r} \right]. \tag{7.16}$$

7.2.2 Blood Flow in an Artery

For the blood with density ρ, the flow is governed by the Navier-Stokes equation and the continuity equation. The material of the wall with density ρ_ω also satisfies the continuity equation. Further, the components of the distortion of the arterial material can be written in a Newtonian sense, in terms of the components of the stress tensor.

$$\rho_\omega \frac{\partial^2 u_r}{\partial t^2} = \frac{\partial \sigma_{rr}}{\partial r} + \frac{\partial \sigma_{rz}}{\partial z} + \frac{\sigma_{rr} - \sigma_{\theta\theta}}{r},$$

$$\rho_\omega \frac{\partial^2 u_z}{\partial t^2} = \frac{\partial \sigma_{rz}}{\partial r} + \frac{\partial \sigma_{zz}}{\partial z} + \frac{\sigma_{rz}}{r}. \tag{7.17}$$

Let us assume that the kinematic viscosity η is constant, and, that the blood is an incompressible Newtonian fluid, $\nabla . \mathbf{v} = 0$. Let us also assume that the fluid flow be linear, i.e., we drop nonlinear terms: $(\mathbf{v}.\nabla)\mathbf{v} \ll \frac{\eta}{\rho} \nabla^2 \mathbf{v}$. There is a complete azimuthal symmetry. With these, we can write the equations for the blood flow as follows:

$$\rho \frac{\partial v_r}{\partial t} = -\frac{\partial P}{\partial r} + \frac{\eta}{\rho} \frac{\partial}{\partial r} \left[\frac{1}{r} \frac{\partial}{\partial r} (r v_r) + \frac{\partial^2 v_r}{\partial z^2} \right],$$

$$\rho \frac{\partial v_z}{\partial t} = -\frac{\partial P}{\partial z} + \frac{\eta}{\rho} \left[\frac{1}{r} \frac{\partial}{\partial r} \left(r \frac{\partial v_z}{\partial r} \right) + \frac{\partial^2 v_z}{\partial z^2} \right],$$

$$\frac{\partial v_z}{\partial z} + \frac{1}{r} \frac{\partial}{\partial r} (r v_r) = 0. \tag{7.18}$$

These equations imply that $\nabla^2 P = 0$. We can solve this using the separation of variables: $P = R(r)Z(z)$. We write with an arbitrary separation constant, $-k^2$,

$$-k^2 = \frac{1}{Z} \frac{d^2 Z}{dz^2} = -\frac{1}{R} \frac{1}{r} \frac{\partial}{\partial r} \left(r \frac{\partial R}{\partial r} \right). \tag{7.19}$$

This yields the respective solutions as e^{ikz} and $J_0(ikr)$. Pressure has the spatiotemporal dependence,

$$P(r, z, t) = A J_0(ikr) e^{ikz} e^{i\omega t}. \tag{7.20}$$

Substituting in (7.18), we look for $v_z(r, z, t)$ of the form, $v_z(r) e^{ikz} e^{i\omega t}$. We see that the radial part, $v_z(r)$ satisfies

$$\frac{d^2 v_z}{dr^2} + \frac{1}{r} \frac{dv_z}{dr} - \gamma^2 v_z = -\frac{ikA}{\eta} J_0(ikr) \tag{7.21}$$

where $\gamma^2 = k^2 + i\omega\rho/\eta$. The homogeneous solution is $v_z^h = J_0(i\gamma r)$, and the particular solution is $v_z^P = B J_0(ikr)$. Thus,

$$v_z(r, z, t) = \left[\frac{kA}{\omega\rho} J_0(ikr) + C J_0(i\gamma r) \right] e^{ikz} e^{i\omega t}. \tag{7.22}$$

Similarly,

$$v_r(r, z, t) = \left[\frac{kA}{\omega\rho} J_1(ikr) + D J_1(i\gamma r) \right] e^{ikz} e^{i\omega t}. \tag{7.23}$$

Thus, the Navier-Stokes and continuity equations are solved. This describes motion of fluid in artery, subject to approximations enumerated above.

Let us discuss the equations of motion for the walls before we discuss boundary conditions. For simplification, we assume that the material is purely elastic, obeying Hooke's law. It is also assumed that solid is incompressible, so that $\nabla \cdot \mathbf{u} = 0$. The components of stress tensor are

$$\sigma_{rr} = Y \frac{\partial u_r}{\partial r},$$

$$\sigma_{\theta\theta} = Y \frac{u_r}{r},$$

$$\sigma_{rz} = \frac{Y}{2} \left[\frac{\partial u_r}{\partial z} + \frac{\partial u_z}{\partial r} \right]. \tag{7.24}$$

The equations describing the components of deformation (displacement) are as follows:

$$\rho_w \frac{\partial^2 u_r}{\partial t^2} = Y \frac{\partial^2 u_r}{\partial r^2} + 2Y \left[\frac{\partial^2 u_r}{\partial z^2} + \frac{\partial^2 u_z}{\partial r \partial z} \right] + \frac{Y}{r} \frac{\partial u_r}{\partial r} - \frac{Y}{r^2} u_r,$$

$$\rho_w \frac{\partial^2 u_z}{\partial t^2} = 2Y \left[\frac{\partial^2 u_r}{\partial r^2} + \frac{\partial^2 u_z}{\partial r \partial z} \right] + Y \frac{\partial^2 u_z}{\partial z^2} + \frac{Y}{2r} \left[\frac{\partial u_r}{\partial r} + \frac{\partial u_z}{\partial r} \right]. \tag{7.25}$$

These are coupled equations and are solved numerically, in general. However, certain simplifications may allow analytical tractability.

When a pulse moves down the artery, the artery will expand or contract. We assume that the motion is largely in the radial direction and that the motion in z-direction is much lesser. So, $u_z \ll u_r$. For this reason, ignoring u_z, we have the equation for u_r:

$$\frac{\partial^2 u_r}{\partial r^2} + \frac{1}{r} \frac{\partial u_r}{\partial r} + \left[\frac{\omega^2 \rho_\omega}{Y} - 2Yk^2 - \frac{1}{r^2} \right] u_r = 0. \tag{7.26}$$

The solution is

$$u_r(r, z, t) = F J_1(i\Gamma r) e^{i(kz+\omega t)} \tag{7.27}$$

with $\gamma^2 = 2Yk^2 - \omega^2 \rho_\omega / Y$. Boundary conditions imposed on the fluid are at $r = 0$:

$$v_r = 0, \qquad \frac{\partial v_z}{\partial r} = 0 \tag{7.28}$$

because the fluid may not flow away from the centre. These conditions are satisfied automatically by the governing equations. The other boundary condition have to be applied to the inner surface of the artery.

The relative motion between the fluid and wall must vanish at the surface, the corresponding boundary conditions at $r = r_{\text{in}}$ are

$$v_r = \frac{\partial u_r}{\partial t}, \qquad v_z = \frac{\partial u_z}{\partial t}. \tag{7.29}$$

At the inner surface, stresses must be varying continuously. Shear stress along a surface on the inner surface of the artery exerted by the wall is

$$\sigma_{rz} = Y \left(\frac{\partial u_r}{\partial z} + \frac{\partial u_z}{\partial r} \right)_{r=r_{\text{in}}} \tag{7.30}$$

The stress exerted by the fluid is

$$\sigma'_{rz} = \eta \left(\frac{\partial v_r}{\partial z} + \frac{\partial v_z}{\partial r} \right). \tag{7.31}$$

We must have $\sigma_{rz} = \sigma'_{rz}$ at $r = r_{\text{in}}$. Similarly, for the stresses along the radial direction equal, ensuing

$$-P + 2\eta \frac{\partial v_r}{\partial r} = 2Y \frac{\partial u_r}{\partial r}. \tag{7.32}$$

At the outer boundary, the stresses exerted by the wall must be continuous with those exerted by the surrounding medium (zero in the case of vacuum).

Generally, the deformation of the artery is not small compared to its radius. The inner and outer radii are functions of time, and this has to be incorporated in the boundary conditions. Let us consider a large artery (say, aorta) of radius a with deformation, ξ, thence $r_{\text{in}} = a + \xi$, with $\xi/a \ll 1$. Now we apply the b.c., and assume that the constant, A is available from the initial conditions. We would like to determine C, D, F in terms of A, using the equations for $v_r, v_z, and u_r$.

From (8.19), ignoring $u_z, v_z = 0$ at $r = a$. This gives

$$C = -\frac{k}{\omega\rho} A \frac{J_0(ika)}{J_0(i\gamma a)}. \tag{7.33}$$

Equation (7.29) further yields a relation between two ther constants:

$$\frac{k}{\omega\rho} A J_1(ika) + D J_1(i\gamma a) = i\omega F J_1(\Gamma a). \tag{7.34}$$

The third relation comes from the condition on stresses (7.32),

$$A \left[-J_0(ika) + \frac{2ik^2\eta}{\omega\rho} J_1'(ika) \right] + i\gamma D J_1'(i\gamma a) = 2iYF\Gamma J_1'(\Gamma a) \tag{7.35}$$

where prime denotes differentiation w.r.t. argument.

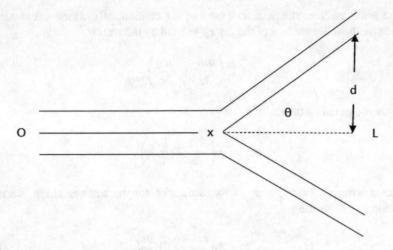

Fig. 7.2 A symmetric branching of an artery is shown here. The point where the branching occurs (x) and the angle θ determine power dissipation (see text)

Exercise 7.1 During *Embryogenesis*, Thompson [2] suggested that the angle of branching of artery is determined by minimizing the energy dissipation. For a symmetric bifurcation (Fig. 7.2) in an artery, with inlet (outlet) flux and radius of branches $Q_{1(2)}$ and $r_{1(2)}$, determine the origin x and bifurcation angle θ for given values of $Q_1/Q_2, r_1/r_2$. Assume that there is no change in kinetic energy from inlet to outlet for a steady flow.

Exercise 7.2 Consider an inviscid, incompressible fluid of density ρ flowing in an elastic tube of cross-sectional area $A(x)$ at a distance x. Let $p(x, t)$ be pressure and velocity parallel to the tube axis be $u(x, t)$. The net force in the positive x-direction on the volume element is seen to be

$$pA + p_0 \frac{\partial A}{\partial x} \Delta x - \left[pA + \frac{\partial (pA)}{\partial x} \Delta x \right] = -\frac{\partial}{\partial x} [(p - p_0) A] \Delta x.$$

Show that the equation of continuity is $\partial A/\partial t + \partial (Au)/\partial x = 0$.

Further, assume that pressure-radius relation be [100]

$$p - p_0 = \frac{Yh}{r_0} \left[1 - \sqrt{\frac{A_0}{A}} \right]$$

where $A = \pi r^2$, $A_0 = \pi r_0^2$. Y is the Young's modulus and h is the thickness of the tube. By linearizing the above equations, show that pressure follows the equation [131],

$$\partial^2 p / c^2 \partial t^2 = \partial^2 p / \partial x^2 \quad \text{with} \quad c^2 = Y h / 2\rho r_0.$$

Apply the Young formula for c^2 to the flow of blood through thoracic aorta to find c for dogs. The mean pressure is 100 mm Hg, $Y \approx 4.3 \times 10^6$ dynes/cm^2, $h/a \approx 0.105$, $\rho \approx 1.06$ g/cc (the experimental value [75] is about 5 cm/s).

Water Waves

> The depths of the sea are very silent: in striking contrast to the
> noisiness of the land ... there is noise where the waves break on
> the shore ... everything is in favour of a noiseless motion.
>
> —William Henry Bragg

Waves in water are most striking, beautiful and intriguing. They are also related closely with our day to day lives. In this chapter, after recollecting some definitions, we set ourselves the task of describing waves and wakes, their analogies in plasmas. At different levels, there are sources [42, 52, 65] which we have used.

Let us recall some important concepts and definitions:

1. *Streamline*: Lines drawn parallel to the velocity field at any instant are called streamlines.
2. *Steady and unsteady flow*: If the velocity field is constant at any point, the fluid flow is said to be steady. A steady flow implies that the pattern of streamlines is also constant. On the contrary, for an unsteady flow, pattern of streamlines changes with time. If an elemental arclength along the streamline is denoted by $ds = (dx, dy, dz)$ and the three components of velocity are (u, v, w), along a streamline,

$$\frac{dx}{u} = \frac{dy}{v} = \frac{dz}{w}. \tag{8.1}$$

Integrating (8.1), we obtain the equation of a streamline.
3. *Irrotational flow*: If the individual fluid particles do not rotate, we have

$$\boldsymbol{\omega} = 0 \text{ which implies that } \mathbf{v} = \nabla \phi \tag{8.2}$$

where ϕ is a velocity potential, the flow thus described is a potential flow or an irrotational flow.

4. *Incompressible flow*: If the density of the fluid particles does not change along the fluid motion, it can be considered constant through out the volume for all times, then $\partial\rho/\partial t = 0$, and we get the condition for incompressible flow:

$$\nabla.\mathbf{v} = 0. \tag{8.3}$$

5. *Incompressible, irrotational flow*: Eqs. (8.2) and (8.3) combine to give

$$\nabla^2\phi = 0, \tag{8.4}$$

Laplace equation for the velocity potential. Immediately, we can write (1.22) in terms of ϕ:

$$\nabla\left[\frac{\partial\phi}{\partial t} + \frac{1}{2}\nabla\phi^2 + \frac{p}{\rho}\right] = \mathbf{f}. \tag{8.5}$$

6. *Stream function*: For two-dimensional flow, let $\mathbf{v}(x, z)$ be the velocity field for surface water waves. The continuity equation is that for an incompressible flow:

$$\frac{\partial u}{\partial x} + \frac{\partial w}{\partial z} = 0. \tag{8.6}$$

Let us denote the stream function by $\psi(x, z, t)$ and define it by

$$u = \frac{\partial\psi}{\partial z}, \quad w = -\frac{\partial\psi}{\partial x}. \tag{8.7}$$

Recalling (8.2), we get relationship between velocity potential and stream function:

$$\frac{\partial\phi}{\partial x} = \frac{\partial\psi}{\partial z}, \quad \frac{\partial\phi}{\partial z} = -\frac{\partial\psi}{\partial x}. \tag{8.8}$$

This pair of equations implies the Laplace equation for the stream function,

$$\nabla^2\psi = 0. \tag{8.9}$$

For a steady flow, the stream function is time-independent:

$$d\psi = \frac{\partial\psi}{\partial x}dx + \frac{\partial\psi}{\partial z}dz = -wdx + udz. \tag{8.10}$$

If $d\psi = 0$ along a streamline, the instantaneous streamline is given by $\psi =$ constant. On the other hand, equipotential curves for the velocity are given by ϕ = constant. Since

$$\frac{\partial \psi}{\partial x}\frac{\partial \phi}{\partial x} + \frac{\partial \psi}{\partial z}\frac{\partial \phi}{\partial z} = \frac{\partial \psi}{\partial x}\frac{\partial \psi}{\partial z} + \frac{\partial \psi}{\partial z}\left(-\frac{\partial \psi}{\partial x}\right) = 0. \tag{8.11}$$

This condition simply states that ψ's and ϕ's trace orthogonal curves. We can then define function $V = \phi + i\psi$ of a complex variable, $Z = x + iz$, and the (8.7) become Cauchy-Riemann conditions.

7. *Gravity waves* Water waves in gravitational force ($\mathbf{f} = -g\hat{z}$) satisfy

$$\nabla\left[\frac{\partial \phi}{\partial t} + \frac{1}{2}\nabla\phi^2 + \frac{p}{\rho} + gz\right] = 0. \tag{8.12}$$

On integration,

$$\frac{\partial \phi}{\partial t} + \frac{1}{2}\nabla\phi^2 + \frac{p - p_0}{\rho} + gz = C(t). \tag{8.13}$$

Re-defining potential as $\phi + \int \left(C(t) + \frac{p_0}{\rho}\right) dt$, we get

$$\frac{\partial \phi}{\partial t} + \frac{1}{2}\nabla\phi^2 + \frac{p}{\rho} + gz = 0. \tag{8.14}$$

Boundary Conditions
Laplace equation (8.4) and the equation of motion (8.14) are subjected to boundary conditions according to the problem. At the upper end of a water layer, we have the free surface and at the lower end is the horizontal bed. For the moment, we are neglecting the surface tension between the liquid (water) and the external fluid.

Let us write the free surface as $z = \eta(x, y, t)$ where $\eta = 0$ is the equilibrium condi7tion of the surface. The fluid velocity component

$$w = \frac{dz}{dt} = \frac{\partial \eta}{\partial x}u + \frac{\partial \eta}{\partial y}v + \frac{\partial \eta}{\partial t} \tag{8.15}$$

is to be evaluated at the surface. Using (8.2), we get the kinematic boundary condition at $z = \eta(x, y, t)$ (partial derivatives are written as subscripts):

$$\phi_z = \eta_x\phi_x + \eta_y\phi_y + \eta_t. \tag{8.16}$$

We assume that the both water and fluid are at the same atmospheric pressure, p_a. At $z = \eta$, (8.14) yields the dynamic boundary condition,

$$\phi_t + \frac{1}{2}[(\phi_x)^2 + (\phi_y)^2 + (\phi_z)^2] + g\eta = 0. \tag{8.17}$$

where we have hidden p_a in the definition of velocity, $\mathbf{v} = \nabla(\phi + p_a/\rho)$. Differentiating (8.17) w.r.t. x, we get

$$u_t + uu_x + vv_x + ww_x + g\eta_x = 0. \tag{8.18}$$

At the horizontal bed, there is the rigid fixed boundary where normal velocity of water is zero:

$$w = \phi_z = 0 \text{ at } z = -h \tag{8.19}$$

(u, v are non-zero due to viscous effects).

Statement of the Problem
The water surface is $z = \eta(x, y, t)$ with $\eta = 0$ as equilibrium position. Gravity waves on water can be studied in terms of Euler equations for an irrotational flow of an incompressible, inviscid fluid with a free surface. The relevant equations are as follows.

$$\nabla^2\phi = 0, \quad (-h < z < \eta(x, y, t)), \tag{8.20}$$

$$\phi_z = \eta_x\phi_x + \eta_y\phi_y + \eta_z\phi_z, \quad (z = \eta(x, y, t)), \tag{8.21}$$

$$\phi_t + \frac{1}{2}[(\phi_x)^2 + (\phi_y)^2 + (\phi_z)^2] + g\eta = 0, \quad (z = \eta(x, y, t)), \tag{8.22}$$

$$\phi_z = 0, \quad (z = -h). \tag{8.23}$$

8.1   Small-Amplitude Surface Gravity Waves

We have already set up the equations to describe water waves whose amplitude is small compared to the wavelength, and the equilibrium position is at rest. Equation (8.16) gives the free surface boundary condition (b. c.) which can be linearized as follows. We know η is small, say of $O(\epsilon)$, and initially the velocity is zero, which means that all the components of velocities will also be small. of order $O(\epsilon)$. This makes the first two terms on the R.H.S. of (8.16) of order $O(\epsilon^2)$. Thus,

$$\phi_z = \eta_t, \quad \text{to order } O(\epsilon). \tag{8.24}$$

Similarly, (8.17) simply reduces on linearization of $z = \eta$ about $\eta = 0$ to

$$\phi_t + g\eta = 0, \quad \text{to order } O(\epsilon). \tag{8.25}$$

We can again utilize the fact that η is small and expand ϕ_z in a Taylor series around $z = 0$ to show that it can be calculated to order $O(\epsilon)$ just at $z = 0$ instead of $z = \eta$:

$$\phi_z(\eta) = \phi_z(\eta = 0) + \eta \phi_{zz}(\eta = 0), \tag{8.26}$$

second term is of order $O(\epsilon^2)$.

We can differentiate (8.25) w.r.t. time and use ((8.24) and (8.26)):

$$\phi_{tt} + g\eta_t = \phi_{tt} + g\phi_z = 0 \quad \text{at } z = 0. \tag{8.27}$$

With this, we have converted a nonlinear free boundary problem to a linear fixed boundary problem for the velocity potential:

$$\nabla^2 \phi = 0, \quad -h < z < 0, \tag{8.28}$$

$$\phi_{tt} + g\phi_z = 0, \quad z = 0, \tag{8.29}$$

$$\phi_z = 0, \quad z = -h. \tag{8.30}$$

Let us simply consider waves propagating along x-direction and assume that they are uniform in y-direction, this makes all the quantities independent of y. It is natural to asume that the waves are sinusoidal, thus we look for a separable solution of the Laplace equation:

$$\phi(x, z, t) := q(x, z) \sin(kx - \omega t). \tag{8.31}$$

Laplace equation becomes

$$\left(\frac{\partial^2 q}{\partial x^2} + \frac{\partial^2 q}{\partial z^2} - k^2 q \right) \sin(kx - \omega t) + 2k \frac{\partial q}{\partial x} \cos(kx - \omega t) = 0. \tag{8.32}$$

By Weierstrass-Lindemann theorem on algebraic independence of exponential function, coefficients of cosine and sine have to be zero independently. This implies that

$$\frac{\partial q}{\partial x} = 0, \tag{8.33}$$

q is independent of x. The coefficient of sine becomes then

$$\frac{\partial^2 q}{\partial z^2} - k^2 q = 0 \tag{8.34}$$

which gives the general solution, $q(z) = Ae^{kz} + Be^{-kz}$. Thus, the potential is

$$\phi(x, z, t) = (Ae^{kz} + Be^{-kz}) \sin(kx - \omega t). \tag{8.35}$$

To determine the constants, let us use the fixed b.c.,

$$\frac{\partial \phi}{\partial z}\bigg|_{z=-h} = 0, \text{ implying } B = Ae^{-2kh}. \tag{8.36}$$

Re-arranging the terms, the potential is

$$\phi(x, z, t) = 2Ae^{-kh} \cosh[k(z + h)] \sin(kx - \omega t). \tag{8.37}$$

From (8.24), we see that

$$\frac{\partial \eta}{\partial t} = \frac{\partial \phi}{\partial z}\bigg|_{z=\eta}$$

$$= 2Ake^{-kh} \sinh[k(\eta + h)] \sin(kx - \omega t). \tag{8.38}$$

Solving this by recalling that $\eta = 0$ at $t = 0$,

$$\eta(t) = a \cos(kx - \omega t) \tag{8.39}$$

where

$$a = \frac{2Ak}{\omega} e^{-kh} \sinh(kh). \tag{8.40}$$

We have therefore arrived at the solution for the potential,

$$\phi(x, z, t) = \frac{a\omega}{k \sinh(kh)} \cosh[k(z + h)] \sin(kx - \omega t). \tag{8.41}$$

Substitution of this solution in (8.27) gives the dispersion relation:

$$\omega^2 = gk \tanh(kh). \tag{8.42}$$

From this, the phase velocity, v_p and the group velocity, v_g follow:

$$v_p = \frac{\omega}{k} = \sqrt{\frac{g\lambda}{2\pi} \tanh\left(\frac{2\pi h}{\lambda}\right)}, \tag{8.43}$$

$$v_g = \frac{d\omega}{dk} = \frac{v_p}{2}\left[1 + \frac{kh}{\sinh(2kh)}\right]. \tag{8.44}$$

8.1.1 Linear Shallow-Water Waves

Shallow-water waves will satisfy dispersion relation when $h/\lambda \ll 1$, thus $kh \ll 1$. In this limit,

$$\tanh kh \equiv kh - \frac{(kh)^3}{3}, \tag{8.45}$$

thus,

$$\omega = c_0 k \left[1 - \frac{(kh)^2}{6}\right], \tag{8.46}$$

where $c_0 = \sqrt{gh}$. In the linear limit, $\omega = c_0 k$. The phase and group velocities can be found with this relation. This is a very good approximation as can be checked by considering $h < 0.07\lambda$ in which case the phase velocity is

$$v_p = \sqrt{gh}[1 - (4\pi^2/6)(h/\lambda)^2]$$
$$= \sqrt{gh}[1 - 0.03] \sim 0.97c_0,$$

just 3% away from linear dispersion relation.

8.1.2 Deep-Water Waves

These waves are defined in the limit $\lambda \sim h$ or $h/\lambda \gg 1$. The dispersion relation now takes the approximate form,

$$\omega^2 = gk \tanh(kh) \equiv gk, \quad \text{thus} \quad \omega = \sqrt{gk}. \tag{8.47}$$

The phase velocity is

$$v_p = \sqrt{\frac{g}{k}}, \tag{8.48}$$

clearly, there is dispersion as $v_p \propto \sqrt{\lambda}$. With $g = 9.8 \text{ ms}^{-2}$, $v_p \sim 1.25\sqrt{\lambda}$. For waves in the sea, $50 \text{ m} < h < 150 \text{ m}$, phase velocity is in the range $9 - 150$ m/s.

8.1.3 Effect of Surface Tension

Suppose the pressure in water is p and the atmospheric pressure is p_a. The free surface element experiences a surface tension σ per unit length. With R as the radius

of curvature of the curved surface, the additional force due to surface tension is due
to the pressure

$$\delta p = \frac{\sigma}{R} = -\sigma \frac{\partial^2 \eta}{\partial z^2}. \tag{8.49}$$

The second derivative above is proportional to $(-k^2)$ for a harmonic plane wave,
and thus the dispersion relation is modified to the leading order to

$$\omega^2 = gk + \frac{\sigma}{\rho} k^3. \tag{8.50}$$

When k increases, gravity waves are smoothly transformed to capillary waves with
frequency,

$$\omega = \sqrt{\frac{\sigma}{\rho}} k^{3/2}. \tag{8.51}$$

For water, this takes place at $k \sim 10$ cm^{-1} or a wavelength of 1 cm.
From (8.50), phase velocity is

$$v_p^2 = \frac{g}{k} + \frac{\sigma k}{\rho}. \tag{8.52}$$

Surface water waves exhibit strong dispersion—that is, v_p decreases (increases)
as k increases for small k ($k > k_0 = \sqrt{\frac{g\rho}{\sigma}}$); for water, with $\sigma = 0.072$ Nm^{-2},
$k_0 \sim 3$ cm^{-1}. Minimum value of the phase velocity correspond to $k = k_0$, for
water, it is about 2.5 cm/s.
Combining (8.47) and (8.50), the dispersion relation for water layer of finite
depth is

$$\omega^2 = \left(gk + \frac{\sigma}{\rho} k^3 \right) \tanh(kh). \tag{8.53}$$

When water layer is shallow ($kh \ll 1$), (8.53) becomes

$$\omega^2 = hk^2 \left(g + \frac{\sigma}{\rho} k^2 \right) \left(1 - \frac{k^2 h^2}{3} \right). \tag{8.54}$$

From this, we have two cases:

$$\omega^2 = gkh, \quad \text{(no dispersion)} \tag{8.55}$$

$$\omega^2 = hk^2 g \left(1 - \frac{k^2 h^2}{3} + \frac{\sigma}{\rho g} k^2 \right), \quad \text{(weak dispersion)}. \tag{8.56}$$

8.2 Waves from a Local Pulsed Source

Consider all familiar ripples created by a small stone dropped in a lake. The localized source can be considered as a two-dimensional Dirac delta distribution whose Fourier components are known. Ripples represent the temporal evolution of these Fourier components. Wave pattern consists of circular waves, rings extending upto a certain distance r_0. Considering all the circles into rings in such a way that each ring contains at least several humps, a ring of thickness Δr looks like a circular wave packet. We know that a wave packet propagates with a group velocity,

$$v_g = \frac{\partial \omega}{\partial k}. \tag{8.57}$$

For gravity waves, since $\omega^2 = gk$,

$$v_g = \frac{1}{2}\sqrt{\frac{g}{k}}, \quad v_p = \sqrt{\frac{g}{k}}, \tag{8.58}$$

thus group velocity is half the phase velocity. The distance covered by the wave packet with wave number k in time t is

$$r = v_g t = \frac{t}{2}\sqrt{\frac{g}{k}}, \tag{8.59}$$

or,

$$k = \frac{gt^2}{4r^2}. \tag{8.60}$$

This relation shows that at a fixed point, r, k increases with time. This means that the wave packets are continuously renewed by higher and higher k values. These packets all run with group velocities at each point, r, equal to r/t.

On the other hand, if we take a snapshot, i.e. fix t, we see that at the centre, wave number is very large, decreasing rapidly with increasing r. When $kr \sim 1$, wave pattern disappears; at large distances, there are no waves. We get estimate of the visible boundary of wave structure,

$$r_0 \sim \frac{gt^2}{4}; \tag{8.61}$$

this front expands with constant acceleration, of order g.

Recall that $v_p = 2v_g$—hence crests run twice as fast as the groups of waves and after reaching the boundary of the circle of radius r_0, they disappear.

What about the phase, $\varphi(= \omega t - kr)$ of the wave-pattern ? Phase is defined by the relations:

$$\omega = \frac{\partial \varphi}{\partial t}, \quad k = -\frac{\partial \varphi}{\partial r}. \tag{8.62}$$

Using $\omega = \sqrt{gk}$, we get on using (8.60),

$$\varphi = -\frac{gt^2}{4} \int \frac{dr}{r^2} = \frac{gt^2}{4r}. \tag{8.63}$$

So, at large distances from centre, there can be no waves: the change of phase there is $< \pi$ and no crest can be observed.

8.3   Cerenkov Emission and Kelvin Wakes

We have seen beautiful waves produced by a moving boat or ship or duck. This is simply an instance of Cerenkov radiation by a body moving with velocity greater than the phase velocity. This is mainly because of the dispersive nature of gravity waves.

Cerenkov emission arises as a superposition of elementary waves produced by a moving body. Circular rings are enveloped by Mach lines of constant phase. We see that the angle of emission is given in terms of the speed of the moving body, v_0 and the phase velocity, v_p:

$$\cos \theta = \frac{v_p}{v_0}. \tag{8.64}$$

But,

$$\cos \theta = \frac{k_z}{k}. \tag{8.65}$$

With these,

$$\frac{v_p}{v_0} = \frac{k_z}{k} \quad \text{which implies that} \quad v_p k = \omega = k_z v_0, \tag{8.66}$$

which is the Cerenkov relation. In the moving frame, wave is seen as a steady state pattern.

Fig. 8.1 When a body moves on a water surface with a speed v_0, it generates circular waves, as shown here

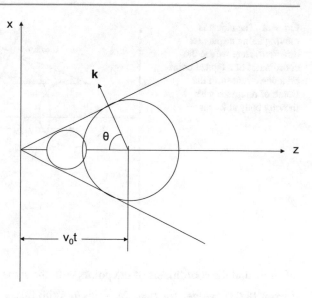

Consider Cerenkov radiation produced in a dispersive medium—emission of gravity waves produced by a body moving on a deep water surface. These care produced by a series of small perturbations, each perturbation generates a lot of circular waves as shown in Fig. 8.1 below.

Let us choose two circular lines which correspond to elementary waves emitted by a moving body at times t and $t + \Delta t$. Assume that these lines correspond to same phase—wavefronts—so the difference in their radii is $v_p \Delta t$. Their centres are separated by the distance, $\Delta z = v_0 \Delta t$. We see that the angle of emission is given by $\cos \theta = v_p / v_0$, as in the dispersionless case. The above relations are as usual.

Let us now study how the wavepackets propagate with their group velocity. We know that the distance between the source and point of observation, $r = v_g t$. This is not equal to $v_p t$ as was in the dispersionless case. Recalling again that $v_g = v_p / 2$ for the gravitational waves, we have

$$r = v_g t = \frac{1}{2} v_p t = \frac{1}{2} v_0 t \cos \theta \qquad (8.67)$$

according to the Cerenkov emission condition. From the Fig. 8.2, it is clear that the coordinates of the point of observation in the frame of reference with moving body at its centre are

$$x = r \sin \theta, \quad z = v_0 t - r \cos \theta. \qquad (8.68)$$

Fig. 8.2 The object is moving along negative x direction. Here we see the coordinates of a typical point on a wavefront from the frame of reference with moving body at its centre

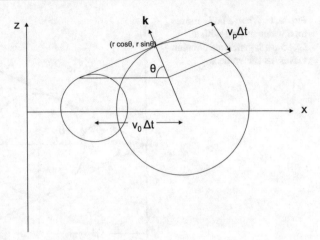

We now find the coordinates of the points with the same phase. In (8.63), substitute r from (8.68), we get for $t = 2\varphi \frac{v_0}{g} \cos\theta$. With this, $r = gt^2/4\varphi = \varphi \frac{v_0^2}{g} \cos^2\theta$. Thus, we get

$$x = \frac{v_0^2 \varphi}{g} \sin\theta \cos^2\theta, \quad \text{and} \tag{8.69}$$

$$z = \frac{v_0^2 \varphi}{g}(2 - \cos^2\theta)\cos\theta \tag{8.70}$$

in terms of phase, φ and the angle, θ between wave vector and z-axis.

As φ changes from 0 to 2π, we obtain the family of lines which describe the net of crests. This net has two types of waves—(1) one travelling sideways from wave central line to periphery. The phase velocity of these waves is small compared to v_0 so θ is not much different from $\pi/2$; (2) second wave looks like distinct fragments of the same progressive wave which propagates with phase velocity equal to the velocity of the moving body: $v_p = v_0$. For this case,

$$k = \frac{g}{v_0^2} \tag{8.71}$$

and this wave propagates together with the body being in resonance with it. As seen in the Fig. 8.3, the wave pattern is confined inside an angle, 2α. The limits on the angle of emission have to be found first and then the values of x, z for those values. Thus, with

$$\left.\frac{\partial x}{\partial \theta}\right|_{\theta_0} = 0 \quad \text{implies} \quad \cos^3\theta_0 - 2\cos\theta_0 \sin^2\theta_0 = 0$$

$$2\sin^2\theta_0 = \cos^2\theta_0 \quad \text{implies} \quad \tan\theta_0 = \frac{1}{\sqrt{2}}, \tag{8.72}$$

Fig. 8.3 The wave pattern is seen here for a typical value of $r = 1.1$. Allowed values of r for which the pattern is confined within 2α, are given by (8.86)

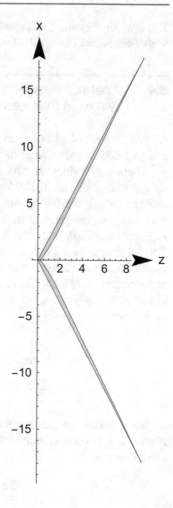

With this, $\cos\theta_0 = \sqrt{2/3}$ and $\sin\theta_0 = 1/\sqrt{3}$. At θ_0, x, z are

$$x(\theta_0) = \frac{v_0^2\varphi}{g}\frac{1}{\sqrt{3}}\frac{2}{3},$$ (8.73)

$$z(\theta_0) = \frac{v_0^2\varphi}{g}\left(2 - \frac{2}{3}\right)\sqrt{\frac{2}{3}},$$ (8.74)

implying thereby that

$$\tan\alpha = \frac{x(\theta_0)}{z(\theta_0)} = \frac{1}{2\sqrt{2}}, \quad \text{or } \alpha = \tan^{-1}\frac{\sqrt{2}}{4} = \tan^{-1}0.3537 \sim 19.5\text{degrees}.$$ (8.75)

Thus, we understand a remarkable observation that the angle made is $39°$ irrespective of the nature of object—that is, the angle is universal!

8.4 Analogy Between Shallow Water Waves and Ion-Acoustic Waves in Plasmas

A weakly ionized plasma can be produced by an external plasma source or by a high-frequency discharge has a low density where electrons follow Boltzmann distribution of velocities with a temperature of few eVs, and ions are very low temperature. For simplicity, let us consider one-dimensional motion of ions. An electric field is created by the difference in shift of electrons and ions, and ions with mass m_i respond to it. The electric field can be calculated by solving the Poisson's equation with density difference between electrons (n_e) and ions (n_i). The ionic density follows a continuity equation. Ions respond slowly, performing oscillations about a mean position. Electrons equilibrate during the compressions and rarefactions of ions. The resulting nonlinear equations can be linearized to result in sound waves—called the ion-acoustic waves. In all, the governing equations are as follows:

$$m_i \frac{dv}{dt} = eE, \quad \frac{\partial n_i}{\partial t} + \frac{\partial}{\partial x}(n_i v) = 0,$$

$$T_e \frac{\partial n_e}{\partial x} = -eEn_e, \quad \frac{\partial E}{\partial x} = 4\pi e(n_i - n_e). \tag{8.76}$$

We consider nonlinear sound waves with very long wavelengths, $k \to zero$. Neglecting dispersion assuming $kd \to 0$ and using quasineutrality condition, the equations are

$$\frac{\partial v}{\partial t} + v\frac{\partial v}{\partial x} + c_s^2 \frac{1}{n}\frac{\partial n}{\partial x} = 0, \tag{8.77}$$

and continuity equation in (8.76). These equations are analogous to equations for shallow water waves with column height, h:

$$\frac{\partial v}{\partial t} + v\frac{\partial v}{\partial x} + g\frac{\partial h}{\partial x} = 0,$$

$$\frac{\partial h}{\partial t} + \frac{\partial(hv)}{\partial x} = 0. \tag{8.78}$$

Linearization of (8.76) is achieved by writing the perturbations n_i' and n_e' over the equilibrium density n_0. After some simple manipulations, one can see that the perturbations obey

$$\frac{\partial^2 n_i'}{\partial t^2} + \omega^2 n_i' = 0, \tag{8.79}$$

with $\omega = c_s k / (1+k^2 d^2)^{1/2}$, where $c_s = k_B T_e / m_i$, $d = \sqrt{k_B T_e / 4\pi e^2 n_0}$. The phase and group velocity are

$$v_p = \frac{c_s}{\sqrt{1 + k^2 d^2}}, \quad v_g = \frac{c_s}{(1 + k^2 d^2)^{3/2}}. \tag{8.80}$$

Waves with short wavelengths (large k's) propagate more slowly than long wave ones. For the case of a pulsed source we have $v_g = r/t$, it gives

$$k = k_0 \sqrt{\left(\frac{c_s t}{r}\right)^{2/3} - 1} \tag{8.81}$$

where $k_0 = 1/d$; this expression holds for $r < c_s t$ (sub-sonic region). From the expression for wavenumber $k = -\partial\varphi/\partial r$,

$$\varphi = -\int k \, dr = k_0 r \left[\left(\frac{c_s t}{r}\right)^{2/3} - 1\right]^{3/2}. \tag{8.82}$$

With the condition of Cerenkov radiation, $\omega/k = v_0 \cos\theta$, r and t can be expressed in terms of φ:

$$r = \varphi d \frac{v_0^3 \cos^3\theta}{c_s^3} \left(1 - \frac{v_0^2}{c_s^2} \cos^2\theta\right)^{-3/2},$$

$$t = \frac{\varphi d}{c_s} \left(1 - \frac{v_0^2}{c_s^2} \cos^2\theta\right)^{-3/2}. \tag{8.83}$$

The wave pattern is obtained in $x - z$ plane by writing $x = r \sin\theta$ and $z = v_0 t - r \cos\theta$. These are easily obtained by using (8.83). This ion-sound pattern produced by a subsonic moving body is seen in Fig. 8.4. The coordinates for the wakes as [41]

$$x = \varphi d \frac{v_0^3}{c_s^3} \sin\theta \cos^3\theta \left(1 - \frac{v_0^2}{c_s^2} \cos^2\theta\right)^{-3/2},$$

$$z = \varphi d \frac{v_0}{c_s} \left(1 - \frac{v_0^2}{c_s^2} \cos^4\theta\right) \left(1 - \frac{v_0^2}{c_s^2} \cos^2\theta\right)^{-3/2}. \tag{8.84}$$

These are valid for subsonic and supersonic cases. For a fixed phase φ, these define the wavefronts. For the wave pattern to be situated inside a certain angle, 2α as in the case of Kelvin wakes in water, $\partial x/\partial\theta = 0$. This gives possible values of $\theta = \theta_0$.

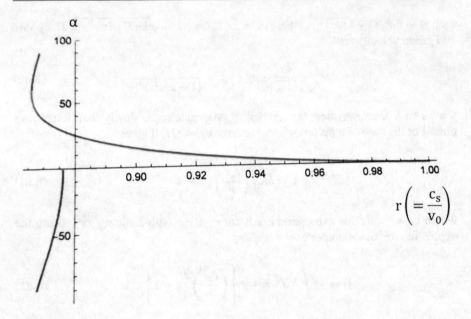

Fig. 8.4 The variation of the values of half-wake angle is seen with the ratio, c_s/v_0. Wakes are Kelvin-like, except that there is a dependence on the speed of the object. Even for the Kelvin wakes in water, a dependence on the speed of the source has been found [95, 97]

These expressions suggest limits on velocity of the moving object, v_0 via the ratio $r^2 = c_s^2/v_0^2$ as

$$\cos\theta_0 = \pm(r^2 \pm r\sqrt{4r^2 - 3})^{1/2}. \tag{8.85}$$

For $r^2 > 3/4$, $\cos\theta_0 = \pm(r^2 + r\sqrt{4r^2 - 3})^{1/2}$. For the other solution, $\cos\theta_0 = \pm(r^2 - r\sqrt{4r^2 - 3})^{1/2}$ to be valid, $r^2 > r\sqrt{4r^2 - 3}$, i.e., $r^2 < 1$. Thus the range of values of r^2 is

$$3/4 < r^2 < 1, \quad \text{or} \quad c_s < v_0 < \sqrt{4/3}c_s, \tag{8.86}$$

a narrow range.

The wake angle α is obtained by Fig. 8.4

$$\tan\alpha = \frac{x(\theta_))}{z(\theta_0)}$$

$$= \frac{(\sqrt{1 - r^2 \mp r\sqrt{4r^2 - 3}})(r^2 \pm r\sqrt{4r^2 - 3})^{3/2}}{r^2 - (r^2 \pm r\sqrt{4r^2 - 3})^2}. \tag{8.87}$$

The half-wake angle, α is plotted for the range of allowed values of the ratio, r.

8.5   Korteweg-de Vries Equation and Solitons

For ion acoustic waves (IAW), weak dispersion can be taken into account by writing the frequency and phase velocity of the harmonic wave:

$$\omega = c_s k(1 - k^2/2k_0^2), \quad v_p = c_0(1 - k^2/2k_0^2), \quad (k_0 = 1/d). \tag{8.88}$$

Analogously, for waves on a shallow water surface, we can write

$$\omega^2 \simeq k^2 g h(1 - k^2 h^2/3 + \sigma k^2/\rho g). \tag{8.89}$$

If surface tension is not important, then

$$\omega/k \simeq gh(1 - k^2 h^2/6). \tag{8.90}$$

Comparing it with phase velocity of ion-acoustic waves, we have

$$c_0 = gh, \quad k_0 = \sqrt{3}/h. \tag{8.91}$$

To take into account weak dispersion in the case of weak nonlinearity, we can see

$$\frac{\partial v}{\partial t} + \frac{c_s^2}{v_p(1 + k^2 d^2)} \frac{\partial v}{\partial x} = 0. \tag{8.92}$$

For IAW propagating in $+x$-direction, we substitute $v_p = c_s/(1 + k^2 d^2)^{1/2}$ with $k^2 d^2 \ll 1$, obtaining the equation for a singe harmonic:

$$\frac{\partial v}{\partial t} + c_s \left(1 - \frac{k^2}{2k_0^2}\right) \frac{\partial v}{\partial x} = 0. \tag{8.93}$$

This can be generalized to a superposition of harmonics by a simple replacement: $k^2 v \rightarrow -\partial^2 v/\partial x^2$. Thus, we obtain

$$\frac{\partial v}{\partial t} + c_s \frac{\partial v}{\partial x} + \frac{c_s k^2}{2k_0^2} \frac{\partial^3 v}{\partial x^3} = 0. \tag{8.94}$$

Introducing a frame of reference moving with c_s, the term $c_s \frac{\partial v}{\partial x}$ disappears from this equation. This equation is linear, describing local ion acceleration under the effect of driving force. If the amplitude of oscillation is not small, we can replace $\partial v/\partial t$ by dv/dt and then local ion acceleration will be in balance with same expression for driving force. The quantity v can be replaced by u which is velocity of media in the steady state frame of reference. The coordinate x is measured in the frame of reference moving with velocity c_0. Finally, thus, we obtain an equation to describe weakly nonlinear waves with weak negative dispersion:

$$\frac{\partial u}{\partial t} + u\frac{\partial u}{\partial x} + \frac{c_0}{2k_0^2}\frac{\partial^3 u}{\partial x^3} = 0, \tag{8.95}$$

the Korteweg-de Vries (KdV) equation.

8.5.1 Periodic Waves, Solitons

Particular solutions correspond to simple progressive waves, $u(x, t) = u(x - ct)$ where c is the phase velocity which is a small increment to c_0 in the laboratory frame. For this progressive wave, $\partial u/\partial t = -c\partial u/\partial x$, which may be substituted in (8.95):

$$-c\frac{\partial u}{\partial x} + u\frac{\partial u}{\partial x} + \frac{c_0}{2k_0^2}\frac{\partial^3 u}{\partial x^3} = 0. \tag{8.96}$$

Integrating over x, and putting the constant of integration to zero, we get a nonlinear ordinary differential equation for u:

$$cu - \frac{u^2}{2} = \frac{c_0}{2k_0^2}\frac{d^2 u}{dx^2}. \tag{8.97}$$

We make a mechanical analogy: (x, u) can be treated as (time, position of a particle). Equation (8.97) is Newton's equation for a particle of mass, $c_0/2k_0^2$:

$$\frac{c_0}{2k_0^2}\frac{d^2 u}{dx^2} = -\frac{\partial W}{\partial u}, \quad W(u) = -\frac{cu^2}{2} + \frac{u^3}{6}. \tag{8.98}$$

The "potential energy" of a periodic wave, has a minimum at c, and it is zero at 0 and $3c$. Periodic wave consists of periodic oscillations of the particle in the potential well. Amplitude of oscillations is related to "energy" of the particle in this well, including kinetic energy. Particle of "low energy" experiences oscillations near the bottom, $u = 2c$. Let us linearize by writing $u = 2c + u'$; the equation for u' is

$$\frac{c_0}{2k_0^2}\frac{d^2 u'}{dx^2} = -cu'. \tag{8.99}$$

For a periodic wave, $u' = u_0' \exp(ik(x - ct))$, where k is related to phase velocity c by $c = c_0 k^2/2k_0^2$. Fluid moves with an average velocity, $2c$. So, the wave propagates with respect to fluid with velocity, $-c$. $\therefore$, in the laboratory frame the wave propagates with phase velocity,

$$v_p = c_0 - c = c_0 - c_0 k^2 / 2k_0^2. \tag{8.100}$$

If we "increase" energy of oscillating particle, amplitude increases, and wave becomes more asymmetrical with crests becoming steeper and more spaced. Finally, consider perturbation with maximum amplitude with zero effective energy. The particle starts motion at $u = 0$, $W = 0$, then it goes to the bottom of the well and climbs up to $u = 3c$, $W = 0$. It experiences backward motion to $u = 0$. Thus, we have only one crest (solitary wave). If we substitute

$$u(x, t) = u_0 \text{sech}^2\{(x - ct)/\Delta\} \tag{8.101}$$

in (8.97), we find that

$$u_0 \Delta^2 = 6c_0 / k_0^2 = \text{constant}, \tag{8.102}$$

i.e., the product of square of thickness and amplitude is a constant. A *soliton* is a supersonic pattern as well as a periodic wave with large amplitude.

8.5.2 Analytical Solution of KdV Equation

KdV equation can be obtained from classical variational principle of least action [35]. Let φ be the velocity potential defined by $u = \partial\varphi/\partial x$. Lagrangian is defined in terms of u:

$$\mathcal{L}(\varphi, \partial\varphi/\partial x, t) = \frac{1}{2} \int \left[\frac{\partial\varphi}{\partial t}\frac{\partial\varphi}{\partial x} + \frac{1}{3}\left(\frac{\partial\varphi}{\partial x}\right)^3 - \frac{c_0}{2x_0^2}\left(\frac{\partial^2\varphi}{\partial x^2}\right)^3 \right] dx. \tag{8.103}$$

The Euler-Lagrange equation is simply the KdV equation. Equation (8.103) implies laws of conservation of energy and linear momentum. Invariance w.r.t. shift along x-axis leads to conservation of momentum,

$$P = \frac{1}{2} \int u^2 \, dx. \tag{8.104}$$

Invariance w.r.t. time implies energy conservation, with energy

$$E_0 = \frac{1}{2} \int \left[-\frac{u^3}{3} + \frac{c_0}{2k_0^2}\left(\frac{\partial u}{\partial x}\right)^2 \right] dx. \tag{8.105}$$

Energy is measured in the frame of reference moving with sound velocity c_0. In laboratory frame, energy is $E = E_0 + c_0 P$. As $u \ll c_0$, $E \sim c_0 P = (c_0/2) \int u^2 \, dx$, there exist an infinite series of integrals of motion.

Analytical theory of the KdV equation uses an analogy with the Schrödinger equation of quantum mechanics with Hamiltonian H:

$$i\partial\psi/\partial t = H\psi. \tag{8.106}$$

If L is a Hermitian operator describing a physical quantity, then its evolution is governed by

$$\frac{dL}{dt} = \frac{\partial L}{\partial t} + i[H, L]. \tag{8.107}$$

If $dL/dt = 0$, L is independent of time. This implies that its eigenvalues λ do not change with time: $L\psi = \lambda\psi$. It turns out that KdV equation can be considered as the steady state condition $dL/dt = 0$ for a specially chosen operator L with special Hamiltonian H. If

$$L = \frac{3c_0}{k_0^2} \frac{\partial^2}{\partial x^2} + u, \tag{8.108}$$

and

$$H = -\frac{2ic_0}{k_0^2} \frac{\partial^3}{\partial x^3} - \frac{i}{2}\left(u\frac{\partial}{\partial x} + \frac{\partial u}{\partial x}\right), \tag{8.109}$$

then by (8.107),

$$\frac{dL}{dt} = 0 = \frac{\partial u}{\partial t} + u\frac{\partial u}{\partial x} + \frac{c_0}{2k_0^2}\frac{\partial^3 u}{\partial x^3}, \tag{8.110}$$

the KdV equation. Eigenvalue equation for L is just Schrödinger equation,

$$-\frac{3c_0}{k_0^2}\frac{\partial^2\psi}{\partial x^2} - u\psi = -\lambda\psi. \tag{8.111}$$

LHS is the sum of kinetic and potential energy, with energy, $-\lambda$ on RHS.

If u decays as $x \to \pm\infty$ rapidly, then positive values of λ correspond to discrete spectrum with bound states, and negative values correspond to an energy continuum. The eigenvalues λ are conserved with time, which implies that when $u(x, t)$ changes according to the KdV equation, then L has constant eigenvalue. We can try to find eigenvalues for solitons. For the soliton profile, $u = u_0^2(x/\Delta)$, (8.111) can be solved exactly, the only eigenvalue is $\lambda = u_0/2$.

Next, the eigenvalue $\lambda = 0$ corresponds to $\psi = \tanh(x/\Delta)$ which does not decay as $x \to \pm\infty$ and thus corresponds to a continuous spectrum.

There are no solitons if $u < 0$ as solitons can be produced by positive perturbations $u > 0$. Any negative perturbation creates at least one soliton, analogous to one-dimensional potential well in quantum mechanics has at least one bound state.

Therefore, total number of solitons together with their amplitude can be found by simply solving (8.111) and not KdV equation. It turns out that (8.111) serves as a tool for exact solution of KdV equation. Asymptotic behaviour of ψ is used for this purpose. If the profile of u decreases sufficiently rapidly as $x \to \pm\infty$, the asymptotic solution is a superposition of bound and scattering eigenfunctions:

$$\psi_n = a_n(t)e^{-\kappa_n x}, \quad \psi_k = a_k(t)e^{-ikx} + b_k(t)e^{ikx}, \tag{8.112}$$

where $\kappa_n = k_0\sqrt{\lambda_n/3c_0}$, $k = k_0\sqrt{-\lambda/3c_0}$, $\lambda_n > 0$ (discrete), $\lambda_n < 0$ (continuous). Time-independent of amplitude in (8.112) can be derived from KdV equation. As $x \to \pm\infty$, the nonlinear term $u\partial u/\partial x$ can be dropped, and linear equation gives the following solutions:

$$a_n(t) = a_n(0)\exp(2c_0\kappa_n^3 t/k_0^2), \quad a_k(t) = a_k(0)\exp(2ic_0\kappa_n^3 t/k_0^2),$$

$$b_n(t) = b_n(0)\exp(-2ic_0\kappa_n^3 t/k_0^2). \tag{8.113}$$

The reflection coefficient is $S_k(t) = b_k(t)/a_k(t)$. From scattering theory, we know that potential $u(x, t)$ can be found if S_k is known as a function of k for the continuous spectrum and the characteristics κ_n and a_n of the discrete spectrum are also known.

Exercise 8.1 Winds blowing along the surface of the ocean generate water currents known as drift currents. The disturbance created on the surface by the wind imparts motion to adjacent layers by friction pressures. Consider the ocean infinitely extended with coordinate system (ξ, η, ζ) attached to a surface point of the Earth. The ξ-axis points to the south, $(\xi - \eta)$ along meridian and parallel axes, and ζ-axis along the outward surface- normal. Assume that density and viscosity are constant.

Magnetohydrodynamics

9

A magnetic field is ubiquitous in the cosmos and one of the intriguing problems in physical cosmology is to seek answer to the question of origin of the all pervasive rotation and magnetic field, and whither the two force-fields came into existence together.

9.1 Equations of Motion

Let us first consider the role of magnetism in astrophysical settings by considering a small volume element $d\tau$ containing electrical charges moving with the mean velocity u. This enables us to define the local charge and current densities respectively as $\rho_e d\tau = \sum_{d\tau} e_i$ and $j d\tau = \sum_{d\tau} e_i u$. The force experienced by a single charge e_i in the presence of an electromagnetic field, (E, B) where E is the electric field and B is the magnetic induction is given by the Lorentz force, $F_i = e_i u \times B/c + e_i E$. The total force acting on the charges, e_i in the volume element, $d\tau$ is given by

$$F = \sum_i \left(e_i E + \frac{1}{c} e_i u \times B \right) = d\tau \left(\rho_e E + \frac{1}{c} j \times B \right) \qquad (9.1)$$

which is the Lorentz force $\rho_e E + \frac{1}{c} j \times B$ per unit volume, acting on an electrically conducting fluid element as a long-range body force to be incorporated in the equation of motion for an inviscid, electrically conducting fluid which then takes the form,

$$\rho \frac{du}{dt} = -\nabla p + \rho \mathbf{g} + \rho_e E + \frac{1}{c} j \times B \qquad (9.2)$$

to be supplemented by Maxwell's equations

$$\nabla.\boldsymbol{B} = 0, \qquad \boldsymbol{B} = \mu\boldsymbol{H},$$

$$\nabla.\boldsymbol{D} = 4\pi\rho_e, \qquad \boldsymbol{D} = \kappa\boldsymbol{E},$$

$$\nabla \times \boldsymbol{H} = \frac{4\pi\boldsymbol{j}}{c} + \frac{1}{c}\frac{\partial\boldsymbol{D}}{\partial t}, \quad \text{(Ampere's law)}$$

$$\nabla \times \boldsymbol{E} = -\frac{1}{c}\frac{\partial\boldsymbol{B}}{\partial t} \quad \text{(Faraday's law)}. \qquad (9.3)$$

Maxwell introduced the displacement current in his electromagnetic set of equations for maintaining the conservation of charge density which turns out to be negligible for good conductors, except for high-frequency phenomena. Adopting a dimensional analysis by introducing a characteristic length-scale, L and time-scale TL/u over which electromagnetic fields have a significant variation, Maxwell's equations may then be expressed as

$$\frac{E}{L} \sim \frac{1}{c}\frac{B}{T} \quad \text{which implies} \quad E \sim B\frac{L}{cT} \sim B\frac{u}{c},$$

$$|\rho_e E| \sim \frac{D}{4\pi L}E \sim \frac{\kappa}{4\pi L}E^2 \sim \frac{\kappa B^2}{4\pi L}\left(\frac{u}{c}\right)^2,$$

$$\left|\frac{\boldsymbol{j}\times\boldsymbol{B}}{c}\right| \sim \frac{B^2}{4\pi\mu L}, \quad \text{so that} \quad \frac{|\rho_e E|}{|\boldsymbol{j}\times\boldsymbol{B}/c|} \simeq (\kappa\mu)\left(\frac{u}{c}\right)^2 \ll 1 \text{ for } u \ll c.$$

$$(9.4)$$

Thus, the electrical force density is negligible compared to the magnetic force density.

Likewise, the displacement current, $|\frac{1}{c}\partial\boldsymbol{D}/\partial t|$ is negligible compared to the conduction current, $|4\pi\boldsymbol{j}/c|$, since

$$\left|\frac{1}{c}\frac{\partial\boldsymbol{D}}{\partial t}\right| \sim \frac{\kappa E}{cT} \sim \frac{\kappa u}{cL}E \sim \frac{\kappa}{L}\left(\frac{u}{c}\right)^2 B \qquad (9.5)$$

and

$$\left|\frac{4\pi\boldsymbol{j}}{c}\right| \sim \frac{4\pi}{c}\frac{c}{4\pi}\text{curl }\boldsymbol{H} \sim \frac{H}{L} \sim \frac{B}{\mu L}. \qquad (9.6)$$

Consequently,

$$\left|\frac{\frac{1}{c}\frac{\partial\boldsymbol{D}}{\partial t}}{\frac{4\pi\boldsymbol{j}}{c}}\right| = (\kappa\mu)\left(\frac{u}{c}\right)^2 \ll 1 \qquad (9.7)$$

for $u \ll c$. The magnetohydrodynamic equations then take the following form:

$$\rho \frac{du}{dt} = -\nabla p + \rho \mathbf{g} + \frac{\mathbf{j} \times \mathbf{B}}{c},$$

$$\nabla . \mathbf{B} = 0, \qquad \mathbf{B} = \mu \mathbf{H}, \quad \text{magnetic induction}$$

$$\nabla . \mathbf{D} = 4\pi \rho_e, \qquad \mathbf{D} = \kappa \mathbf{E}, \quad \text{displacement vector}$$

$$\nabla \times \mathbf{H} = \frac{4\pi}{c} \mathbf{j},$$

$$\nabla \times \mathbf{E} = -\frac{1}{c} \frac{\partial \mathbf{B}}{\partial t}. \tag{9.8}$$

This set needs to be supplemented by Ohm's law in its generalized form for completely ionized as well as partially ionized medium.

9.1.1 Ohm's Law for Completely Ionized Gas

Consider a fully ionized gas consisting of electrons and positive ions. Let v_i be the mean velocity of ions and $(v_i + v_e)$ that of electrons, so that v_e is velocity of the electron gas relative to the ions. then the current density may be defined by $\mathbf{j} = -n_e e v_e$, n_e being the electron number density, arising from the relative motion of electrons, while $\mathbf{j}_i = n_e e v_i$ is the current density due to the ion motion. The bulk fluid velocity, v_b is very nearly the same as the velocity of the ions, since

$$v_b = \frac{n_e m_e v_e + n_i m_i v_i}{n_e m_e + n_i m_i} \simeq v_i \tag{9.9}$$

since $m_e \ll m_i$ and $n_e = n_i$ because of charge neutrality. Tota momentum lost by electrons due to collisions with ions per unit volume per unit time is $n_e m_e v_e / \tau_{ei}$, τ_{ei} being the mean collision time of electrons with ions and as a result, electrons suffer a drag force, $-n_e m_e v_e / \tau_{ei}$.

In order to write down the equations of motion of electron gas, we enumerate the forces accelerating the electrons: gravitational force, $n_e m_e \mathbf{g}$, gradient of partial pressure due to electron gas, $-\nabla p_e$, electromagnetic force, $-n_e (\mathbf{E} + (v_i + v_e) \times \mathbf{B})/c$ and the drag due to collisions with ions, $-n_e m_e v_e / \tau_{ei}$.

Since $m_e \ll m_i$ and $v_i \ll v_e$, we can write $|n_e m_e dv_e/dt| \ll |n_e m_e v_e / \tau_{ei}|$ as the collision time, τ_{ei} is small compared with the characteristic time, $T = L/v_b$ of velocity variations except for the high-frequency phenomena, i.e. the inertial terms are negligible, except when the time-scale of velocity variations is small compared to the collision time. Likewise, the gravitational force term,

$$|n_e m_e \mathbf{g}| \ll |\nabla p_e| \simeq \frac{1}{2}|\nabla p| \sim \frac{1}{2}|\rho \mathbf{g}| \sim |n_i m_i g| \tag{9.10}$$

since $m_e \ll m_i$. Thus the inertial and gravitational forces are negligible because of smallness of the electron mass m_e compared to the ion mass, m_i. These approximations enable us to write the equation of motion for the electrons

$$0 = -\nabla p_e + n_e e \left(E + \frac{(v_i + v_e) \times B}{c} \right) + \frac{n_e m_e v_e}{\tau_{ei}}. \tag{9.11}$$

Since the current density $j = -n_e e v_e$ and the electrical conductivity,

$$\sigma = \frac{e^2 n_e \tau_{ei}}{m_e} = \frac{n_e e^2}{m_e \nu_{ei}}, \tag{9.12}$$

the equation of motion of the electron gas can be expressed as

$$E + \frac{v_i \times B}{c} + \frac{\nabla p_e}{n_e e} = \frac{j \times B}{c n_e e} + \frac{j}{\sigma} \tag{9.13}$$

which is the generalized Ohm's law with $\sigma = n_e e^2 \tau_{ei}/m_e$. Here $E + \frac{v_i \times B}{c}$ represents the electric field measured in a frame of reference moving with velocity v_i, that is approximately the bulk velocity of the gas, and $\nabla p_e/n_e e$ is the effective electric field due to partial pressure of the electron gas and $(j \times B)/c n_e e$ is the Hall term. Thus, Ohm's law can be written as

$$j = \sigma \left(E + \frac{v_i \times B}{c} + \frac{\nabla p_e}{n_e e} - \sigma \frac{j \times B}{c n_e e} \right)$$

$$= \sigma E_{\text{eff}} - \sigma \frac{j \times B}{c n_e e}, \qquad E_{\text{eff}} = E + \frac{v_i \times B}{c} + \frac{\nabla p_e}{n_e e}. \tag{9.14}$$

This can be re-written as

$$j + \sigma \frac{j \times B}{c n_e e} = \sigma E_{\text{eff}}, \text{ or}$$

$$j + \sigma \frac{j \times B}{B} (\omega_e \tau_{ei}) = \sigma E_{\text{eff}}, \tag{9.15}$$

ω_e being the electron gyro-frequency and the ratio of Hall and conduction current terms may be expressed as

$$\frac{\left| \frac{j \times B}{B} (\omega_e \tau_{ei}) \right|}{|j|} \simeq \omega_e \tau_{ei}. \tag{9.16}$$

The Hall effect clearly dominates when the Hall parameter $H = \omega_e \tau_{ei}$ becomes large which occurs for strong magnetic fields, i.e. electron gyrofrequncy far exceeds the electron-ion collision frequency as well as when τ_{ei} is large which will occur for low density rarefied gas. The Hall parameter H far exceeding unity implies that there is free spiralling of electrons before colliding with the ions. Then the currents flowing are greatly influenced by the magnetic field [21].

9.1.2 Generalization of Ohm's Law for Weakly (Partially) Ionized Gas

The formulation was generalized by Cowling [21] to derive the generalized Ohm's law for a weakly ionized gas composed of electrons, ions, and neutrals, i.e. $n_e = n_i \ll n_a$ with $m_e \ll m_i \simeq m_a$. Thus the fraction of neutrals, $f_a = n_a/(n_i + n_a)$ and $1 - f_a = n_i/(n_i + n_a)$ and mass density $\rho = m_e n_e + m_i n_i + m_a n_a$ while the bulk velocity,

$$v_b = \frac{1}{\rho}[m_i n_i (v_i + v_b) + m_e n_e (v_i + v_b + v_e + m_a n_a v_a] \tag{9.17}$$

with ion velocity, $v_b + v_i$, electron velocity, $v_b + v_i + v_e$, neutral velocity, $v_a = v_b - (n_i/n_a)v_a$. Here the total mass density

$$\rho = m_e n_e + m_i n_i + m_a n_a \simeq m_i(n_i + n_a), \quad \text{since } m_i \simeq m_a, \ m_e \ll m_i,$$

$$= \frac{m_a n_a}{f} \quad \text{noting } n_i \ll n_e. \tag{9.18}$$

Define $\omega_e^c = eB/m_e c$, $\omega_i^c = eB/m_i c$, $\kappa_i = 1/2\omega_i^c \tau_{ei}$, $\kappa_e = 1/\omega_e^c \tau_{en}$, τ_{ei} being the ion-electron collision time, τ_{en} being the electron-neutral collision time, τ_{in} the ion-neutral collision time. Define

$$\beta = \frac{\kappa_e}{\kappa_e + \kappa_i} = \frac{1/\omega_e^c \tau_{en}}{1/\omega_e^c \tau_{en} + 1/2\omega_i^c \tau_{ei}}. \tag{9.19}$$

The generalized Ohm's law for a weakly ionized gas of electrons, ions and neutrals may then be written as [21] neglecting displacement current,

$$\frac{j}{\sigma} = E + \frac{v_b \times B}{c} - \frac{j \times B}{cn_e e}(1 - 2f\beta) + \frac{\nabla p_e}{n_e e}(1 - f\beta)$$

$$+ \left(\frac{j \times B}{cn_e e}\right) \times B \cdot \frac{f^2}{B}\left(\frac{1}{\omega_e^c \tau_{ei}} + \frac{1}{\omega_e^c \tau_{en}}\right)^{-1} - \frac{\nabla p_e \times B}{B} f^2 \left(\frac{1}{\omega_e^c \tau_{ei}} + \frac{1}{\omega_e^c \tau_{en}}\right)^{-1}. \tag{9.20}$$

In the limit of $\beta \ll 1$, the generalized Ohm's law, Hall and ambipolar diffusion effect resulting from the presence of neutrals, takes the form, noting $\beta = 2\omega_i^c \tau_{in}/\omega_i^c \tau_{en} = 2(m_e/m_i)(\tau_{in}/\tau_{en}) \ll 1$,

$$\frac{j}{\sigma} = E + \frac{v_b \times B}{c} - \frac{j \times B}{cn_e e} + \frac{\nabla p_e}{n_e e} + \left(\frac{j \times B}{cn_e e}\right) \times B$$

$$\cdot \frac{f^2}{B}\left(\frac{1}{\omega_e^c \tau_{ei}} + \frac{1}{\omega_e^c \tau_{en}}\right)^{-1} - \frac{\nabla p_e \times B}{B} f^2 \left(\frac{1}{\omega_e^c \tau_{ei}} + \frac{1}{\omega_e^c \tau_{en}}\right)^{-1}. \tag{9.21}$$

Note

$$\frac{1}{1 + \frac{1}{2}\frac{\omega_e^c \tau_{en}}{\omega_i^c \tau_{in}}} \simeq \frac{2\omega_i^c \tau_{in}}{\omega_e^c \tau_{en}} \simeq 2\frac{m_e}{m_i}\frac{\tau_{in}}{\tau_{en}} \ll 1 \tag{9.22}$$

and

$$\sigma = \frac{n_e e}{\left[\frac{1}{\omega_e^c \tau_{ei}} + (1 - \beta)\frac{1}{\omega_e^c \tau_{en}}\right]B}. \tag{9.23}$$

Using Maxwell's equations, $\nabla.B = 0$, $\nabla \times E = -(1/c)\partial B/\partial t$, we recover the generalized induction equation,

$$\frac{\partial B}{\partial t} = \text{curl}\,(v_b \times B) - c\,\text{curl}\left\{\frac{\text{curl}\,B \times B}{4\pi n_e e \mu}\right\}(1 - 2f\beta)$$

$$+ \text{curl}\left\{\frac{\text{curl}\,B \times B}{4\pi \rho v_{in}}f\right\} + \eta\nabla^2 B \tag{9.24}$$

with $\eta = c^2/4\pi\mu\sigma$. In this equation representing the temporal evolution of the magnetic induction, B, the first term on the RHS represents the advection of the magnetic flux by fluid motion, the second term stands for the Hall diffusion and the last term arises from ambipolar diffusion in the presence of neutrals. Here we have neglected contribution arising from partial pressure, grad p_e due to electrons, but in the presence of Hall and ambipolar diffusion. In the absence of ambipolar diffusion, the generalized Ohm's law,

$$\frac{j}{\sigma} + \frac{j \times B}{cn_e e} = \mathcal{E}_{\text{eff}} = E + \frac{u \times B}{c} + \frac{\nabla p_e}{n_e e} \tag{9.25}$$

which can be written as $j = \sigma\mathcal{E}_{\text{eff}}$ in the case of $\mathcal{E}_{\text{eff}} \parallel B$ and

$$j = \sigma^I \mathcal{E}_{\text{eff}} + \sigma^{II}\frac{B \times \mathcal{E}_{\text{eff}}}{B} \tag{9.26}$$

for $\mathcal{E}_{\mathrm{eff}} \perp \boldsymbol{B}$. Here

$$\sigma^{I} = \frac{\sigma}{1 + (\omega_e^c \tau_{ei})^2} \quad \text{and} \quad \sigma^{II} = \frac{\sigma \omega_e^c \tau_{ei}}{1 + (\omega_e^c \tau_{ei})^2}. \tag{9.27}$$

When $\omega_e^c \tau_{ei} \ll 1$, we recover $\sigma^{I} \sim \sigma$, and $\sigma^{II} \ll \sigma$ and for $\omega_e^c \tau_{ei} \gg 1$, $\sigma^{I} \sim \frac{\sigma}{(\omega_e^c \tau_{ei} \ll 1)^2}$, $\sigma^{II} \sim \frac{\sigma}{\omega_e^c \tau_{ei}}$ and $\sigma^{II} \gg \sigma^{I}$. In the latter case, the electron gyrofrequency, ω_e^c is large compared to the collision frequency and there is free spiralling of electrons before collision with the ions. Then the electric currents flowing are greatly influenced by the magnetic field.

In summary, there are 21 variables, $\boldsymbol{E}, \boldsymbol{B}, \boldsymbol{D}, \boldsymbol{H}, \boldsymbol{j}, \boldsymbol{u}, p, \rho, t$ for which we have 21 equations, curl $\boldsymbol{E} = -(1/c)\partial \boldsymbol{B}/\partial t$, div $\boldsymbol{B} = 0$, curl $\boldsymbol{H} = 4\pi \boldsymbol{j}/c$, div $\boldsymbol{D} = 4\pi \rho_e$, Ohm's law, and momentum equation, $\boldsymbol{D} = \kappa \boldsymbol{E}$, $\boldsymbol{B} = \mu \boldsymbol{H}$, $p = p(\rho)$ (neglecting displacement current).

9.2 Role of Maxwell Stresses in Magnetohydrodynamics

9.2.1 Lorentz Force

In Lorentz force,

$$\frac{\boldsymbol{j} \times \boldsymbol{B}}{c} = \frac{(\nabla \times \boldsymbol{B}) \times \boldsymbol{B}}{4\pi \mu} = \frac{(\boldsymbol{B}.\nabla)\boldsymbol{B}}{4\pi \mu} - \nabla \left(\frac{B^2}{8\pi \mu} \right). \tag{9.28}$$

The first term on RHS is non-vanishing provided $\boldsymbol{B}$ varies along the direction of $\boldsymbol{B}$ and basically represents the effects of tension $(B^2/4\pi \mu)$ per unit area parallel to the magnetic lines of force while the second term represents the isotropic pressure, $B^2/8\pi \mu$, pr unit area due to the magnetic field. Thus the mechanical force due to the magnetic field is equivalent to a hydrostatic pressure, along with the tension which has the following consequences: Because of longitudinal tension, the line of force tend to shorten themselves and consequently get bent, while on account of the lateral pressure, a bundle of lines of force tend to resist any lateral compression, and as a result of both, on displacement from equilibrium, the resulting force set up tends to produce oscillations about equilibrium.

9.2.2 Hydromagnetic Waves

Consider an infinite, homogeneous, ideal, incompressible fluid pervaded by a uniform magnetic field, $\boldsymbol{B}_0$, undergoing a small perturbation as a result of which a small velocity field, $\boldsymbol{v}$, is set up causing the magnetic field to be altered to $\boldsymbol{B}_0 + \mathbf{b}$.

The governing equations then become

$$\frac{\partial \boldsymbol{B}}{\partial t} = \operatorname{curl}(\boldsymbol{v} \times \boldsymbol{B}), \qquad (\eta = c^2/4\pi\mu\sigma \equiv 0),$$

$$\rho\frac{\partial \boldsymbol{v}}{\partial t} = -\operatorname{grad} p + \rho\boldsymbol{g} + \frac{(\nabla \times \boldsymbol{B}) \times \boldsymbol{B}}{4\pi\mu}. \tag{9.29}$$

Linearizing these equations, we recover

$$\frac{\partial \mathbf{b}}{\partial t} = \operatorname{curl}(\boldsymbol{v} \times \boldsymbol{B}_0),$$

$$\rho\frac{\partial \boldsymbol{v}}{\partial t} = -\operatorname{grad} p + \rho\,\mathbf{grad}\,\psi + \frac{\nabla \times (\mathbf{b} \times \boldsymbol{B}_0)}{4\pi} \tag{9.30}$$

where $\mathbf{g} = -\operatorname{grad}\psi$, $\operatorname{div}\boldsymbol{v} = 0$, and $\operatorname{div}\mathbf{b} = 0$, these foregoing equations reduce to

$$\frac{\partial \mathbf{b}}{\partial t} = (\boldsymbol{B}_0.\nabla)\boldsymbol{v}$$

$$\rho\frac{\partial \boldsymbol{v}}{\partial t} = -\operatorname{grad}\left(p + \frac{\boldsymbol{B}_0.\mathbf{b}}{8\pi\mu} + \rho\psi\right) + \frac{(\boldsymbol{B}.\nabla)\boldsymbol{B}}{4\pi\mu}. \tag{9.31}$$

Taking the divergence of the second equation, we get

$$\nabla^2\left(p + \frac{\boldsymbol{B}_0.\mathbf{b}}{8\pi\mu} + \rho\psi\right) = 0. \tag{9.32}$$

External to the region affected by the disturbance $\mathbf{b} = 0$ and from equilibrium condition, $\operatorname{grad}(p + \rho\psi) = 0$. Thus, $\left(p + \frac{\boldsymbol{B}_0.\mathbf{b}}{8\pi\mu} + \rho\psi\right)$ is a solution of Laplace's equation which is a constant outside a particular domain, and so must be constant everywhere to give

$$\rho\frac{\partial \mathbf{V}}{\partial t} = \frac{1}{4\pi\mu}(\boldsymbol{B}_0.\nabla)\mathbf{b}. \tag{9.33}$$

For the purpose of illustration, if we take $\boldsymbol{B}_0 = B_0\hat{z}$, then the equations become

$$\frac{\partial \mathbf{b}}{\partial t} = B_0\frac{\partial \boldsymbol{v}}{\partial z}, \qquad \frac{\partial \boldsymbol{v}}{\partial t} = \frac{B_0}{4\pi\mu}\frac{\partial \mathbf{b}}{\partial z}. \tag{9.34}$$

Combining these two equations, we recover

$$\frac{\partial^2 \mathbf{b}}{\partial t^2} = \frac{B_0^2}{4\pi\rho\mu}\frac{\partial^2 \mathbf{b}}{\partial z^2}, \qquad \frac{\partial^2 v}{\partial t^2} = \frac{B_0^2}{4\pi\rho\mu}\frac{\partial^2 v}{\partial z^2} \tag{9.35}$$

which are just the wave equations with wave velocity satisfied by v and b travelling
with the wave velocity $v_A = \pm\frac{B_0}{\sqrt{4\pi\rho\mu}}$, called the Alfvèn velocity. Alfvèn waves
are transverse in their characteristic propagation with velocity, v_A along the lines of
force, analogous to the effects of tension in an elastic string permitting transmission
along the string with speed, $(T/\rho)^{1/2}$. The MHD behaviour implies, these are
the sets of waves propagating in opposite directions with velocity $\pm v_A$. Broadly
speaking, the Alfvèn waves measure the degree of interaction between magnetic and
velocity fields. Thus, if $v_A \gg v$, the magnetic field dominates to control the fluid
motion, while in the opposite limit, $v_A \ll v$, the magnetic field has no influence on
the fluid motion.

Alfvèn waves which are transverse in nature just like waves on the string with
tension, $T = B_0^2/4\pi\mu$ travelling with velocity $\pm v_A$ along the lines of force. They
measure the degree of interaction with magnetic and velocity fields; thus, for MHD
behaviour, $v_A \simeq v$ and if $V_A \ll v$, the magnetic field has no influence on the motion
while for $v_A \gg v$, the magnetic field dominates or controls the fluid motion.

For an incompressible, inviscid fluid, the vorticity equation becomes

$$\frac{\partial\omega}{\partial t} = \text{curl } (\boldsymbol{u} \times \boldsymbol{\omega}), \tag{9.36}$$

where $\omega = \nabla \times \boldsymbol{u}$ and when combined with the mass conservation equation, $\partial\rho/\partial t +$
$\text{div } (\rho\boldsymbol{u}) = 0$ takes the form,

$$\frac{d}{dt}\left(\frac{\omega}{\rho}\right) = \left(\frac{\omega}{\rho}.\nabla\right)\boldsymbol{u}. \tag{9.37}$$

Analogously, from the magnetic induction equation for a perfectly conducting fluid
(resistivity, $\eta = c^2/4\pi\mu\sigma = 0$), and the use of mass conservation equation yields
the equation,

$$\frac{d}{dt}\left(\frac{\boldsymbol{B}}{\rho}\right) = \left(\frac{\boldsymbol{B}}{\rho}.\nabla\right)\boldsymbol{u}. \tag{9.38}$$

In both cases, if there is no change of $\boldsymbol{u}$ in the direction of vorticity ω and magnetic
induction $\boldsymbol{B}$, it follows

$$\frac{d}{dt}\left(\frac{\omega}{\rho}\right) = 0, \qquad \frac{d}{dt}\left(\frac{\boldsymbol{B}}{\rho}\right) = 0. \tag{9.39}$$

It is interesting to note that the vorticity equation for an incompressible fluid in the
presence of Lorentz force, $\boldsymbol{j} \times \boldsymbol{B}/c$, becomes, after taking the curl of the momentum
equation,

$$\frac{\partial\omega}{\partial t} = \text{curl } (\boldsymbol{u} \times \boldsymbol{\omega}) + \text{curl}\left\{\frac{\nabla \times \boldsymbol{B} \times \boldsymbol{B}}{4\pi\rho}\right\} + \nu\nabla^2\omega, \tag{9.40}$$

where $v = \mu/\rho$ is kinematic viscosity. Thus, both the velocity and consequently vorticity field get modified by the action of Maxwell stresses.

On the other hand, in the presence of Hall current in the generalized Ohm's law, the induction equation takes the form

$$\frac{\partial B}{\partial t} = \text{curl}\,(u \times \omega) - c\,\text{curl}\left\{\frac{\nabla \times B \times B}{4\pi n_e e}\right\} + \eta\nabla^2 B, \tag{9.41}$$

where $\eta = c^2/4\pi\mu\sigma$ is the resistivity, with σ the electrical conductivity.

For a fully ionized gas with $Zn_i = n_e$ and $\rho = n_i m_i$, we obtain

$$\frac{\partial}{\partial t}\left(\frac{ZeB}{m_i c}\right) = \text{curl}\left(u \times \frac{ZeB}{m_i c}\right) - \text{curl}\left\{\frac{\nabla \times B \times B}{4\pi\rho_i}\right\} + \eta\nabla^2\left(\frac{ZeB}{m_i c}\right). \tag{9.42}$$

Combining the two Eqs. (9.40) and (9.42), we recover

$$\frac{\partial}{\partial t}\left(\omega + \frac{ZeB}{m_i c}\right) = \text{curl}\left(u \times \left(\omega + \frac{ZeB}{m_i c}\right)\right) + \left(v\nabla^2\omega + \eta\nabla^2\left(\frac{ZeB}{m_i c}\right)\right). \tag{9.43}$$

For an ideal fluid with $v = 0$, $\eta = 0$, we obtain

$$\frac{d}{dt}\left(\omega + \frac{ZeB}{m_i c}\right) = 0, \tag{9.44}$$

implying that the magnetovorticity, $\omega_M = \omega + \frac{ZeB}{m_i c}$ is conserved in the absence of diffusive terms.

The induction equation,

$$\frac{\partial B}{\partial t} = \text{curl}\,(u \times \omega) + \eta\nabla^2 B, \tag{9.45}$$

essentially encapsulates the time rate of change o the magnetic field resulting from the rate at which the magnetic flux is transported or convected by fluid motions and the rate at which the lines of force leak or diffuse through the conducting medium. In this flux-conservation equation, the advective term dominates the diffusive term provided the magnetic Reynolds number $R_m = Lv/\eta \gg 1$. The flux conservation equation then becomes

$$\frac{\partial B}{\partial t} = \text{curl}\,(u \times \omega), \tag{9.46}$$

i.e. the temporal field changes are the same as if the magnetic lines of force are constrained, or frozen to move with the material, formally this condition arises if the electrical conductivity, σ is, infinite or equivalently the resistivity is negligible. On the other hand, for the material at rest, the flux conservation equation becomes

$$\frac{\partial \boldsymbol{B}}{\partial t} = \frac{c^2}{4\pi \mu \sigma} \nabla^2 \boldsymbol{B} := \eta \nabla^2 \boldsymbol{B}, \tag{9.47}$$

where $\eta = c^2/4\pi \mu \sigma$ is the magnetic diffusivity. This is basically the diffusion equation indicating that the magnetic field 'leaks' through the material with characteristic decay time,

$$\tau \sim \frac{4\pi \mu \sigma L^2}{c^2}. \tag{9.48}$$

The freezing of the lines of force condition is rarely satisfied in laboratory situations, where the magnetic Reynolds number, $R_m \simeq O(1)$, while in the astrophysical context both the scale-length, L and velocity, v as well as the electrical conductivity, σ are high for R_m to have sufficiently large values. Interestingly, in the laboratory situations, the causal order is that the electrical field intensity, E and electrical conductivity σ determine the current density j which in turn determines the magnetic field, H from curl $H = 4\pi j/c$. In the cosmic context, this order is, indeed, reversed—the magnetic field changes are very nearly given by the flux conservation, (9.46) and the current density then follows from curl $H = 4\pi j/c$ and [div]$j = 0$, and the sole function of the electrical conductivity σ is to determine the very tiny electric field, $E + v \times B/c$ required to produce the current.

In the context of MHD, the induction equation determines behaviour of the magnetic field, once the velocity field v is known, behaviour which depends crucially on whether the magnetic Reynolds number $R_m \gg 1$ or $\ll 1$. It is noteworthy that for incompressible electrically conducting fluid, the governing equations are:

$$\text{div } \boldsymbol{B} = 0,$$

$$\frac{\partial \boldsymbol{B}}{\partial t} + (\boldsymbol{u}.\text{grad}) \boldsymbol{B} = (\boldsymbol{B}.\text{grad}) \boldsymbol{u} + \eta \nabla^2 \boldsymbol{B}, \tag{9.49}$$

from (9.46) which is similar to the vorticity equation for an incompressible fluid (9.40). The analogy between the vorticity and magnetic field was taken to imply that the magnetic lines respond to motion in a similar way to the classical behaviour of vortex lines, i.e. like the magnetic field, vortex lines are also transported by the flow velocity, u and partly diffuse through it.

Note that the foregoing arguments are applicable for an incompressible fluid. In a turbulent medium the particles tend to diffuse apart, in the process lenghthening the vortex lines which move with the fluid. Similarly, starting with a small RMS value of magnetic field will tend to increase as a result of stretching of the lines of force by fluid motion. However, there is one major significant difference between vorticity and magnetic induction, that is while ω and velocity, u, both properties of the fluid are connected by the relation $\omega = \nabla \times u$, except that the fluid motion can be modified by a sufficiently strong magnetic field through the action of Lorentz force in the equation of motion, while the vorticity is a property of the fluid which does not influence the velocity field, u characterizing the fluid motion.

Exercise 9.1 Explain generation of magnetic field when density and temperature gradients inside completely ionized plasma are perpendicular to each other. Show that such a situation leads to generation of seed magnetic field inside the plasma i.e. spontaneous generation of magnetic field when no initial magnetic field is present inside the plasma.

9.3   Force-Free Magnetic Fields and Beltrami Flows

The configurations of force-free magnetic fields, curl $B = \lambda B$ and aligned helical flows, curl $u = \lambda u$ with the vorticity, $\omega = \nabla \times u$ parallel to u. Here the quantities, $j.B = $ curl $B.B$ and $u.$curl $u = u.\omega$ are known as helicities, associated respectively with B and u. Both the aligned helical flows and the force-free magnetic fields likewise satisfy equation of the form, curl $Q = \alpha Q$.

Such aligned helical flows provide a good basis for describing the solar coronal loops, since in the corona gravitational forces are negligible and magnetic fields are believed to be almost force-free [92]. These have been extensively studied in the context of photospheric magnetic fields of the Sun, where the coronal magnetic fields are assumed to be extrapolations of photospheric magnetic fields. This aspect of the so-called Beltrami flows given by curl $u = \lambda u$ simplifies the nonlinear term— one part of $(u.\nabla)u$ term consists of curl $u \times u$ and the other, grad($u^2/2$) which can be absorbed in pressure term. The main effect of rearrangement of the nonlinear term, $(u.\nabla)u$ is that it prohibits the cascade to smaller scales leading to the build-up at larger scales. Such an effect is also applicable to the magnetic field term, $(B.\nabla)B$ arising from the Lorentz force.

The small scale helical turbulence is known to contribute to the build-up of the large-scale magnetic field generated through the so-called alpha effect in the mean field dynamo formulation. Basically, both helical velocity fields and force-free magnetic fields on small and large scales are expected to contribute to the generation of large-scale fields. Of course, the complete problem can only be addressed by invoking both the momentum and magnetic flux equations. In summary, large-scale flows can be generated from small-scale flows, while the large-scale magnetic fields can be generated from small-scale flows and small-scale magnetic fields, for example, by invoking Hall effect and the current helicity.

Batchelor based his arguments on the analogy between the magnetic induction equation,

$$\frac{\partial B}{\partial t} = \text{curl } (u \times B) + \frac{c^2}{4\pi\mu\sigma}\nabla^2 B, \tag{9.50}$$

and the vorticity equation, for an incompressible fluid,

$$\frac{\partial \boldsymbol{\omega}}{\partial t} = \text{curl} \, (\boldsymbol{u} \times \boldsymbol{\omega}) + \nu \nabla^2 \boldsymbol{\omega}. \tag{9.51}$$

Since both $\boldsymbol{\omega}$ and $\boldsymbol{B}$ are solenoidal vectors, Batchelor concluded that the general behaviour of the magnetic field could be inferred from that of vorticity in ordinary hydrodynamics. But there is an important difference in the two equations which was pointed out by Cowling that in the equation for vorticity which is an intrinsic property of the fluid, the velocity field is that in the absence of a magnetic field. Thus, for the analogy between the equations satisfied by $\boldsymbol{B}$ and $\boldsymbol{\omega}$ to hold the velocity field must not be essentially affected or altered by the presence of magnetic field which need not be the case.

The relation, curl $\boldsymbol{B} = \alpha \boldsymbol{B}$ implies that as one travels in the direction of $\boldsymbol{B}$ along a magnetic field line, the neighbouring field lines would curl in a fixed sense around the field line. Thus, a force-field is essentially a twisted field.

9.3.1 Magnetic Field Configurations

(a) The potential magnetic field, $\boldsymbol{B}$ is curl-free and the corresponding, $\boldsymbol{j} = (c/4\pi)\nabla \times \mathbf{H} = 0$ and consequently, the Lorentz force, $\boldsymbol{j} \times \boldsymbol{B}/c$ vanishes.

(b) The force-free field also entails the vanishing of the Lorentz force, but occurs when the current density $\boldsymbol{j} \parallel \boldsymbol{B}$. Noting that div $\boldsymbol{B} = 0$, the Lorentz force $\boldsymbol{j} \times \boldsymbol{B}/c = \text{curl} \, \boldsymbol{B} \times \boldsymbol{B}/4\pi \mu$ vanishes when curl $\boldsymbol{B}$ is parallel to $\boldsymbol{B}$, or equivalently,

$$\nabla \times \boldsymbol{B} = \alpha \boldsymbol{B}. \tag{9.52}$$

Taking curl of both the sides gives curl $\boldsymbol{B} = -\nabla^2 \boldsymbol{B} = \text{curl} \, (\alpha \boldsymbol{B})$, using div $\boldsymbol{B} = 0$, or $-\nabla^2 \boldsymbol{B} = \alpha \text{curl} \, \boldsymbol{B} = \alpha^2 \boldsymbol{B}$, for a constant value of α, i.e.

$$(\nabla^2 + \alpha^2)\boldsymbol{B} = 0. \tag{9.53}$$

Equally, taking divergence of both sides of (9.52) gives

$$\text{div} \, (\alpha \boldsymbol{B}) = (\boldsymbol{B}.\nabla)\alpha = 0, \tag{9.54}$$

or α is constant along a magnetic field line, i.e. $\boldsymbol{B}$ lies on surfaces of constant α. It may be noted that if such a surface is closed, it cannot in general be simply connected. An example of a force-free configuration is a torus or a doughnut-shaped surface confined by external pressure.

Exercise 9.2 Consider a control volume V bound by a surface S. A magnetic field configuration is called as vacuum magnetic field when it is produced entirely by currents outside V. Show that such a magnetic field, subject to prescribed boundary conditions on S, is uniquely determined by Laplace equation for a scalar function, ψ such that $\mathbf{B} = \nabla \psi$ where $\mathbf{B}$ is the vacuum magnetic field inside V.

Exercise 9.3 Using variational principle show that for a control volume V, described in the Exercise 9.2, the vacuum magnetic field configuration forms the minimum energy state configuration.

9.3.2 Hydromagnetic Waves

No magnetic field of finite energy can be force-free everywhere, since if $j \times B = 0$ everywhere, the magnetic energy, $\int \mathbf{r}.(j \times B)dr = 0$. Thus, $\int (B^2/8\pi\mu)d\tau = 0$ and consequently $B = 0$ everywhere. But in reality if the Alfvèn speed is comparable with the sound speed, $v_A \simeq c_s$ we cannot neglect the effects of compressibility or density variations. Hence for large $\beta = P_{gas}/(B^2/8\pi\mu)$, or equivalently, $c_s \geq v_A$, the changes in the density on account of changes in the magnetic field are quickly relieved through the sound waves. In the event the value of β is close to unity, there is a strong interaction between the effects of compressibility and of the magnetic fields. This can be analyzed in the framework of a linearized theory, wherein the unperturbed state will assume the pressure p_0 and density ρ_0 to be uniform and ignore the pressure of gravitational force. Similarly the initial magnetic field, $\mathbf{B}$ is also taken to be uniform in the z-direction, $B_0 \hat{z}$.

The passage of a wave changes the undisturbed p_0 and ρ_0 at any point by p_1 and ρ_1, where $p_1 = \gamma p_0(\rho_1/\rho_0) = c_s^2 \rho_1$. The perturbed equations in the linear order then become

$$\frac{\partial \rho_1}{\partial t} = -\rho_0 \nabla.\mathbf{u}, \qquad p_1 = c_s^2 \rho_1,$$

$$\rho_0 \frac{\partial \mathbf{u}}{\partial t} = -\nabla p_1 + \frac{1}{4\pi\mu}\left\{ B_0 \frac{\partial \mathbf{b}}{\partial z} - \nabla(\mathbf{B_0.b}) \right\},$$

$$\frac{\partial \mathbf{b}}{\partial t} = \nabla \times (\mathbf{u} \times \mathbf{B_0}) = B_0 \frac{\partial \mathbf{u}}{\partial z} - \mathbf{B_0}(\nabla.\mathbf{u}) \tag{9.55}$$

where $v_A = B_0^2/4\pi\mu\rho_0$, $\mathbf{B} = \mathbf{B_0} + \mathbf{b}$, $\mathbf{u} = 0 + \mathbf{u}$. After elimination we recover the equation

$$\frac{\partial^2 \mathbf{u}}{\partial t^2} = c_s^2 \nabla(\nabla.\mathbf{u}) + v_A^2 \left\{ \frac{\partial}{\partial z}\left(\frac{\partial \mathbf{u}}{\partial z} - \hat{z}\nabla.\mathbf{u} \right) \right.$$

$$\left. - \nabla\left(\frac{\partial u_z}{\partial z} - \nabla.\mathbf{u} \right) \right\}. \tag{9.56}$$

Taking the z-component and divergence of this equation, we get

$$\frac{\partial^2 u_z}{\partial t^2} = c_s^2 \frac{\partial}{\partial z}(\nabla.\boldsymbol{u}),$$

$$\frac{\partial^2}{\partial t^2}(\nabla.\boldsymbol{u}) = c_s^2 \nabla^2(\nabla.\boldsymbol{u}) + v_A^2 \nabla^2 \left(\nabla.\boldsymbol{u} - \frac{\partial u_z}{\partial z}\right). \tag{9.57}$$

Trivially, if the fluid is incompressible, i.e. div $\boldsymbol{u} = 0$, then $u_z = 0$, and $p_1 = \rho_1 = 0$ to reduce the equation to

$$\frac{\partial^2 \boldsymbol{u}}{\partial t^2} = v_A^2 \frac{\partial^2 \boldsymbol{u}}{\partial z^2} \tag{9.58}$$

which corresponds to Alfvèn waves propagating along with the field lines. Here compressibility does not affect the waves which are wholly transverse twist or shear waves. On the other hand, if div $\boldsymbol{u} \neq 0$, elimination of u_z between the equations gives

$$\frac{\partial^4}{\partial t^4}(\nabla.\boldsymbol{u}) - (v_A^2 + c_s^2)\frac{\partial^2}{\partial t^2}(\nabla^2 \text{div } \boldsymbol{u}) + v_A^2 c_s^2 \frac{\partial^2}{\partial z^2}(\nabla^2 \text{div } \boldsymbol{u}) = 0. \tag{9.59}$$

For a plane wave for which the perturbed quantities are all proportional to $\exp(i\omega(t - \boldsymbol{n}.\boldsymbol{r}/u))$, where u is the phase velocity and $\boldsymbol{n}$ is the unit vector representing its direction. Suppose the wave normal makes an angle θ with $\boldsymbol{B}_0$, then the resulting dispersion relation becomes

$$\omega^4 - k^2(v_A^2 + c_s^2)\omega^2 + k^4 v_A^2 c_s^2 \cos^2 \theta = 0. \tag{9.60}$$

Thus, for any angle θ, there are two wave speeds, u corresponding to *fast* waves for which $u > c_s$ and v_A and *slow* waves corresponding to $u < c_s$ and v_A. The solution of the foregoing equation in the limit of $\omega_I \ll \omega_A$ is $\omega^2 = \omega_A^2(1 \pm \omega_I/\omega_A)$, so that the Coriolis force produces a small splitting of the Alfvèn wave frequency.

9.4 Ferraro's Law of Isorotation

The magnetic rigidity imparted by the electrically conducting material as a result of large electrical conductivity (or, negligible resistivity, $\eta = c^2/4\pi\mu\sigma$) imposes certain conditions on the possible magnetic field configuration that can exist inside the body. One example of this relates to the non-uniform rotation of a star possessing a magnetic field; for example, it is observed that the Sun rotates faster at the equator than in the polar regions. Let us illustrate this for the case of our Sun which may be considered to be made of infinitely conducting electrical fluid in its interior with a steady axisymmetric motion inside the body. Suppose the axisymmetric

$(\partial/\partial\theta = 0)$ magnetic field configuration has no azimuthal magnetic field, i.e. $B_\theta = 0$ in cylindrical geometry (w, θ, z) for a meridional magnetic field. Then, the Ampere's law in the absence of displacement current gives $\nabla \times H = 4\pi j/c$, imply that the current density j has only the θ-component in cylindrical geometry while the Ohm's law gives $j/\sigma = E + u \times B/c$.

Taking the curl of this equation, we recover the θ-component of this equation in the form,

$$\frac{\partial}{\partial z}\left(\frac{u_\theta}{w}B_z\right) + \frac{\partial}{\partial \tilde{w}}\left(\frac{u_\theta}{w}\tilde{w}B_z\right), \tag{9.61}$$

while div $B = 0$ implies

$$\frac{1}{w}\frac{\partial}{\partial z}(\tilde{w}B_z) + \frac{1}{\tilde{w}}\frac{\partial}{\partial z}(\tilde{w}B_{\tilde{w}}) = 0. \tag{9.62}$$

Combining the last two equations, we obtain

$$B_z\frac{\partial}{\partial z}\left(\frac{u_\theta}{\tilde{w}}\right) + B_{\tilde{w}}\frac{\partial}{\partial \tilde{w}}\left(\frac{u_\theta}{w}\right). \tag{9.63}$$

Writing $u_\theta = \tilde{w}\Omega$, Ω being the angular velocity about the axis, the equation reduces to

$$(B.\text{grad})\Omega = 0, \tag{9.64}$$

i.e., the angular velocity Ω is constant along a magnetic field line in a spherical body, or equivalently, different shells rotate at different rates containing their magnetic fields.

9.5   Magnetic Reconnections and Formation of Current Sheets

Current sheets can be formed as a result of interaction of topologically distinct parts of a magnetic field configuration giving rise to active regions on the solar surface, as also at the interface between two flux-tubes (Fig. 9.1). The oppositely directed magnetic field lines of strength, B_0, frozen in the plasma may be transported towards one another at a speed, v_A by a converging flow. When they enter the diffusive region with dimensions (L, ℓ) as indicated in the figure, they could get reconnected at a neutral point. The field lines are carried into the diffusive region from the sides with velocity v_i and out through the top and bottom with speed $v_0 \simeq v_A$.

A typical scenario for formation of current sheet: such a sheet may be created either due to a magnetic field or non-equilibrium phenomenon. The reconnection itself can develop in several different ways: (i) spontaneously generated by a resistive instability such as the tearing mode; (ii) may be driven externally where separate flux-tubes of opposite polarity are pushed together resulting in the region

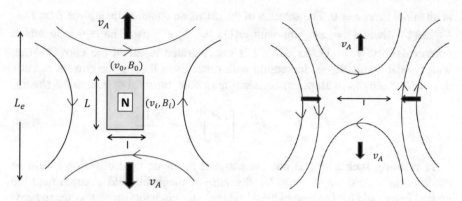

Fig. 9.1 Schematic diagram showing formation of current sheets. Reconnection can occur at a neutral point, **N** after the field lines enter the diffusive region

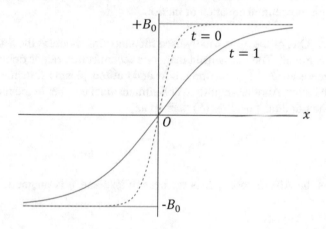

Fig. 9.2 Diffusion of magnetic field

close to neutral lines; (iii) could form locally by a sudden enhancement of the resistivity at some location on account of a process such as ambipolar diffusion.

As an example, consider the diffusion equation $\partial \boldsymbol{B}/\partial t = \eta \nabla^2 \boldsymbol{B}$ (η, resistivity), which dictates magnetic field variations over length-scale, ℓ that are destroyed over the diffusion time of order, $t_{\text{diff}} \sim \ell^2/\eta$.

Consider the diffusion of a unidirectional magnetic field, $\boldsymbol{B}_0 = B_0(x, t)\hat{y}$ with (Fig. 9.2)

$$B_0\,(x, 0) = \begin{cases} +B_0\,, & x > 0, \\ -B_0\,, & x < 0, \end{cases} \qquad (9.65)$$

at an initial time, $t = 0$. The solution of the diffusion equation then gives $B(x, t) = \frac{2B_0}{\sqrt{\pi}} \text{erf}(\xi)$ where $\xi = x/\sqrt{4\eta t}$ with $\text{erf}(\xi) = \int_0^\xi e^{-u^2} du$. The region in which current density, $j = (1/4\pi)dB/dx$ is concentrated is called the current sheet, with typical width, $4\sqrt{\eta t}$, increasing with time. Even though the current density, $j_z = (c/4\pi)dB/dx$ at all points is changing in time, the total current in the sheet,

$$J = \int_{-\infty}^{\infty} j_z dx = \left[\frac{cB}{4\pi}\right]_{-\infty}^{\infty} = \frac{2cB_0}{4\pi}. \tag{9.66}$$

In practice, such a simple one-dimensional magnetic field diffusion would be modified in several ways. Thus, the decrease of magnetic field strength near the neutral line would lead to inward magnetic pressure gradients driving a flow towards the neutral lines from the sides and outwards along them. This naturally requires the inclusion of advective term, curl ($\boldsymbol{u} \times \boldsymbol{B}$) in the induction equation giving rise to a coupling with momentum equation of motion.

Exercise 9.4 One of the most widely used reconnection model is the Sweet-Parker reconnection model. The normalized magnetic reconnection rate is defined as $R = V_{in}/V_{out}$ where V_{in} and V_{out} are the inflow and outflow plasma velocities as shown in the figure below. Another important non-dimensional number in the reconnection physics is the Lundquist number (S) defined as

$$S = \frac{V_A \Delta}{\eta}, \tag{9.67}$$

where V_A is the Alfven speed, Δ is the system size and η is magnetic diffusivity (Fig. 9.3).

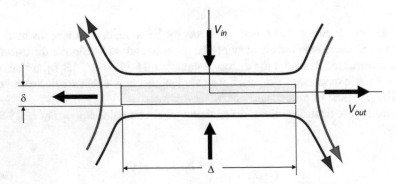

Fig. 9.3 Sweet-Parker model

Using Sweet-Parker reconnection model, show that in steady-state reconnection rate is given by

$$R \simeq S^{-1/2} \tag{9.68}$$

by following steps given below:

- Use mass conservation to obtain relation between V_{in} and V_{out}.
- Use energy conservation to show that $V_{out} = V_A$.
- Write expression for electric field inside the current sheet by considering only resistive term of the Ohm's law. The scaling for current density can be obtained from Ampere's law.
- Equate above electric field with that outside the current sheet where resistive term from Ohm's law is neglected.
- Rearrange the terms to obtain the expression for R in terms of S.

Tensor Virial Theorem and Applications

10

Here we follow the treatment outlined by Chandrasekhar [58, 59]. There are four broad classes of forces in Nature: Gravitational, Electromagnetic have a long-range in character and Strong, Weak interactions which are short-range. Let us consider a simple illustration of an ideal fluid (with no viscosity, no resistivity and no thermal diffusivity) at pressure, p, density, ρ, velocity, v, in magnetic field, B, under the influence of gravitational force. Assuming for a perfect ideal gas the internal energy $E = p/(\gamma - 1)$ where $\gamma = C_p/C_V$ is assumed to remain constant. The equation of motion for the gas neglecting the rotational forces but including the Lorentz force may be written as

$$\rho \frac{d\boldsymbol{u}}{dt} = -\nabla p + \rho \mathbf{g} + \frac{\boldsymbol{j} \times \boldsymbol{B}}{c}. \tag{10.1}$$

Here the gravitational acceleration, $\mathbf{g}$ is expressed in terms of gravitational potential, ϕ as $-\nabla \phi$. Supplementing the Maxwell's equations in the form div $\boldsymbol{B} = 0$, $\boldsymbol{B}$ being the magnetic induction and $\boldsymbol{j} = (c/4\pi)$curl $\boldsymbol{H}$ where $\boldsymbol{H}$ is the magnetic intensity related by $\boldsymbol{B} = \mu \boldsymbol{H}$, where μ is the magnetic permeability and the displacement current is neglected. Following the treatment of Chandrasekhar [59], the equation of motion in the suffix notation may be written as

$$\rho \frac{du_i}{dt} = -\frac{\partial p}{\partial x_i} - \rho \frac{\partial \Phi}{\partial x_i} - \frac{\partial}{\partial x_i}\left(\frac{B^2}{8\pi\mu}\right) + \frac{1}{4\pi\mu}\frac{\partial}{\partial x_j}(B_i B_j) \tag{10.2}$$

where we have used $\partial B_i/\partial x_i = 0$. The scalar gravitational potential is

$$\Phi(\mathbf{r}) = -G \int \frac{\rho(\mathbf{r}')d^3 r'}{|\mathbf{r} - \mathbf{r}'|} \tag{10.3}$$

© The Author(s), under exclusive license to Springer Nature Switzerland AG 2022
S. R. Jain et al., *A Primer on Fluid Mechanics with Applications*,
https://doi.org/10.1007/978-3-031-20487-6_10

where $d^3r' = dx_1' dx_2' dx_3'$ is the volume occupied by a fluid element, $\rho d^3r'$. We introduce some notation where summation over repeated indices is assumed:

$$\Phi_{ik}(r) = -G \int d^3r' \frac{\rho(r')(x_i - x_i')(x_k - x_k')}{|r - r'|^3} \tag{10.4}$$

so that $\Phi_{ii}(r) = \phi(r)$. and

$$|r - r'|^2 = \sqrt{(x_1 - x_1')^2 + (x_2 - x_2')^2 + (x_3 - x_3')^2}. \tag{10.5}$$

Further, introduce a symmetric tensor,

$$\Omega_{ik} = -\frac{1}{2}G \int \int \frac{\rho(r)\rho(r')(x_i - x_i')(x_k - x_k') d^3r d^3r'}{|r - r'|^3}$$

$$\text{where } \Omega_{ii} = -\frac{1}{2}G \int \int \frac{\rho(r)\rho(r') d^3r d^3r'}{|r - r'|}, \tag{10.6}$$

is simply the gravitational potential energy.

Lemma 10.1 *Establish*

$$\Omega_{ik} = -\int \rho(\mathbf{r}) x_i \frac{\partial \Phi}{\partial x_k} d^3r = -\int \rho(\mathbf{r}) x_k \frac{\partial \Phi}{\partial x_i} d^3r. \tag{10.7}$$

Proof From the form of Φ,

$$\frac{\partial \Phi}{\partial x_k} = G \int \frac{\rho(r')(x_k - x_k') d^3r'}{|r - r'|^3} \tag{10.8}$$

using (10.5). Similarly,

$$\frac{\partial \Phi}{\partial x_i} = G \int \frac{\rho(r')(x_i - x_i') d^3r'}{|r - r'|^3}. \tag{10.9}$$

Consequently,

$$\int \rho(\mathbf{r}) x_k \frac{\partial \Phi}{\partial x_i} d^3r = G \int \int \frac{\rho(r)\rho(r')(x_i - x_i') x_k d^3r d^3r'}{|r - r'|^3}$$

$$= G \int \int \frac{\rho(r)\rho(r')(x_i' - x_i) x_k' d^3r d^3r'}{|r - r'|^3} \tag{10.10}$$

and

$$\int \rho(\mathbf{r}) x_i \frac{\partial \Phi}{\partial x_k} d^3 r = G \int \int \frac{\rho(r)\rho(r')(x_k - x_k')x_i d^3 r d^3 r'}{|r - r'|^3}$$

$$= G \int \int \frac{\rho(r)\rho(r')(x_k' - x_k)x_i' d^3 r d^3 r'}{|r - r'|^3}. \tag{10.11}$$

Thus, interchanging $\mathbf{r}$ and $\mathbf{r}'$,

$$\text{LHS} = \frac{1}{2} \text{ sum of the two RHS} = -\Omega_{ik}, \tag{10.12}$$

hence the proof.

Let us introduce the following definitions: Kinetic energy tensor,

$$T_{ik} = \frac{1}{2} \int \rho v_i v_k d^3 r = T_{ki},$$

$$\text{and } T = T_{ii} = \frac{1}{2} \int \rho v_i^2 d^3 r. \tag{10.13}$$

Magnetic energy tensor is defined through

$$M_{ik} = \frac{1}{8\pi\mu} \int B_i B_k d^3 r = M_{ki}, \text{ and } M = M_{ii} = \frac{1}{8\pi\mu} \int B_i^2 d^3 r.$$

Internal energy

$$E = \frac{1}{(\gamma - 1)} \int p d^3 r = \frac{\Gamma}{\gamma - 1}. \tag{10.14}$$

The inertia tensor is given by

$$I_{ik} = \int \rho x_i x_k d^3 r, \tag{10.15}$$

with moment of inertia,

$$I = \int \rho |x|^2 d^3 r. \tag{10.16}$$

Back to the derivation, multiply the equation of motion by x_k and integrate over the whole volume to get

$$\int \rho x_k \frac{dv_i}{dt} d^3 r = \int \rho x_k \frac{d^2 x_i}{dt^2} d^3 r$$

$$= \int \rho \frac{d}{dt} \left(x_k \frac{dx_i}{dt} \right) d^3 r - \int \rho \frac{dx_k}{dt} \frac{dx_i}{dt} d^3 r$$

$$= \int \rho \frac{d}{dt} \left(x_k \frac{dx_i}{dt} \right) d^3 r - 2T_{ik}. \tag{10.17}$$

The first and third terms on the RHS of (10.2) become

$$-\int_\tau x_k \frac{\partial}{\partial x_i} \left(p + \frac{B^2}{8\pi\mu} \right) d^3 r = -\int_S \left(p + \frac{B^2}{8\pi\mu} \right) x_k dS_i + \delta_{ik} \int_V \left(p + \frac{B^2}{8\pi\mu} \right) d^3 r$$

$$= -\int \left(p + \frac{B^2}{8\pi\mu} \right) x_k dS_i + \delta_{ik}[(\gamma - 1)E + M]. \tag{10.18}$$

Gravitational term on the RHS of (10.2) is

$$-\int \rho x_k \frac{\partial \Phi}{\partial x_i} d^3 r = \Omega_{ik}. \tag{10.19}$$

The magnetic term on the RHS of (10.2) gives

$$\frac{1}{4\pi\mu} \int x_k \frac{\partial}{\partial x_j} (B_i B_j) d^3 r = \frac{1}{4\pi\mu} \int_S x_k B_i B_j dS_i - \frac{\delta_{ij}}{4\pi\mu} \int B_i B_j d^3 r$$

$$= \frac{1}{4\pi\mu} \int_S x_k B_i B_j dS_i - 2M_{ik}. \tag{10.20}$$

Collecting all terms, we recover

$$\int_V \rho \frac{d}{dt} \left(x_k \frac{dx_i}{dt} \right) d^3 r = 2T_{ik} + \delta_{ik}[(\gamma - 1)E + M] + \Omega_{ik} - 2M_{ik}$$

$$+ \frac{1}{8\pi\mu} \int x_k [2B_i B_j dS_i - |B|^2 dS_i] - \int_S p x_k dS_i]. \tag{10.21}$$

Assume that p and $|B|$ decrease sufficiently rapidly for the surface terms to be negligible at larger $\mathbf{r}$. RHS is symmetrical in i and k, and, hence the LHS must also be symmetrical in i and k. Thus,

$$\int \rho \frac{d}{dt}\left(x_k \frac{dx_i}{dt}\right) d^3 r = 2T_{ik} + \delta_{ik}[(\gamma - 1)E + M] + \Omega_{ik} - 2M_{ik}. \qquad (10.22)$$

The symmetry characteristic in i and k of the RHS implies

$$\int \rho \frac{d}{dt}\left(x_k \frac{dx_i}{dt} - \int \rho \frac{d}{dt} x_i \frac{dx_k}{dt}\right) d^3 r = 0. \qquad (10.23)$$

The mass conservation implies $\int d/dt(\rho d^3 r) = 0$ and yields

$$\frac{d}{dt} \int \rho \left(x_k \frac{dx_i}{dt} - \int \rho \frac{d}{dt} x_i \frac{dx_k}{dt}\right) d^3 r = 0, \qquad (10.24)$$

expressing the constancy of total angular momentum of the system. We also recover

$$\text{LHS} = \frac{1}{2} \int \rho \frac{d}{dt}\left(x_i \frac{dx_k}{dt} + x_k \frac{dx_i}{dt}\right) d^3 r = \frac{1}{2} \int \rho \frac{d^2}{dt^2}(x_i x_k) d^3 r$$

$$= \frac{d^2}{dt^2} \int \frac{1}{2} \rho x_i x_k d^3 r = \frac{1}{2} \frac{d^2 I_{ik}}{dt^2}. \qquad (10.25)$$

We then have the tensor Virial theorem in the form,

$$\frac{1}{2} \frac{d^2 I_{ik}}{dt^2} = 2T_{ik} - 2M_{ik} + \Omega_{ik} + \delta_{ik}[(\gamma - 1)E + M]. \qquad (10.26)$$

If $i \neq k$, then δ_{ik} and we recover

$$\frac{1}{2} \frac{d^2 I_{ik}}{dt^2} = 2T_{ik} - 2M_{ik} + \Omega_{ik} \qquad (10.27)$$

while for $i = k$, we obtain the scalar virial theorem:

$$\frac{1}{2} \frac{d^2 I}{dt^2} = 2T - M + \Omega + 3(\gamma - 1)E. \qquad (10.28)$$

In steady state, we recover $2T + M + 3(\gamma - 1)E = -\Omega$. The total energy U of the system is $U = T + M + \Omega + E$ which is $-T - (3\gamma - 4)E$. In the absence of any motion, $U = -(3\gamma - 4)E$. We would then expect explosive instability if the total energy, $U > 0$ which is realized only if $\gamma < 4/3$ which is the condition for instability of the system in the absence of any internal motion, since then a system would have enough energy to expand to infinity.

Note that the scalar Virial Theorem implies that is supported against gravity partly by internal thermal pressure, partly by the magnetic pressure and partly by the kinetic pressure. Further, in the absence of any motion ($T = 0$), we recover

$$U = \frac{\gamma - 4/3}{\gamma - 1}(M + \Omega) = \frac{\gamma - 4/3}{\gamma - 1}(|M - \Omega|). \tag{10.29}$$

Thus, in a stable situation, the total energy, U increases through negative values towards zero as the magnetic energy, $\mathcal{M}$ increases to approach $|\Omega|$. Consequently, the stability of a system diminishes with the increasing magnetic field.

One direct application of the Virial theorem is that we can set the Virial limit to various magnitudes of energies, thus $M \leq -\Omega$, $2T \leq -\Omega$, $E \leq -\Omega/3(\gamma - 1)$.

10.1 Neutron Stars

A simple example is the upper limit of a typical neutron star of mass, $M \simeq M_\odot = 2 \times 10^{33}$ g and radius $R = 10$ km with $\Omega = -\frac{3}{5}GM^2/R$, assuming uniform internal density

A star is supported against gravity, partly by thermal pressure, Partly by magnetic pressure, and partly by kinetic pressure. A neutron star has radius, $R \sim 10^6$ cm, and mass, $M \sim 2 \times 10^{33}$ g. It is interesting to calculate maximum magnetic field strength possible inside such a star.

We know that

$$\Omega = \frac{3}{5}\frac{GM^2}{R} = 1.6 \times 10^{53} \text{ ergs.} \tag{10.30}$$

The magnetic energy should be smaller than this, i.e.,

$$\frac{B^2}{8\pi\mu}\frac{4}{3}\pi R^3 < \frac{3}{5}\frac{GM^2}{R},$$

which implies that

$$B < 10^{18}G. \tag{10.31}$$

To calculate the maximum speed of rotation that a neutron star can have, we need to use the fact that the rotational kinetic energy will be lesser than gravitational energy. A simple calculation yields a staggering number, 50,000 rad/s! It can rotate 10^3 times in a second, whereas the Sun only 10^{-6}.

Total energy of the system in the absence of motion is

$$U = \frac{\gamma - 4/3}{\gamma - 1}(M - |\Omega|). \tag{10.32}$$

Thus, in a stable situation, U increases through negative values towards zero as the magnetic energy approaches the gravitational energy $|\Omega|$. Thus, the stability of a star decreases as the magnetic field increases.

Stability Problems in Hydrodynamics and Hydromagnetics

11

> *What is really important is the fact not that there are a few patterns, but that those patterns are capable of almost endless variations.*
>
> —Jorge Luis Borges

11.1 Linear Stability Analysis

Equilibrium requires that the forces acting on the particle vanish: this can happen at the top of an inverted parabola or at the well. In order to test the stability, we note that on being slightly displaced from the equilibrium position z_0, the force experienced is zero there, i.e., $F(z_0) = 0$. Force can be Taylor-expanded:

$$F(z_0 + dz) = F(z_0) + \left.\frac{dF}{dz}\right|_{z_0} dz + O(z^2)$$

$$= \left.\frac{dF}{dz}\right|_{z_0} dz + O(z^2). \tag{11.1}$$

In terms of the potential, $\phi(z)$, $F(z) = -\partial\phi/\partial z$ at $z = z_0$. In terms of the potential, $F(z_0 + dz) = -(\partial^2\phi/\partial z^2)|_{z=z_0}\, dz$. If $(\partial^2\phi/\partial z^2)|_{z=z_0} > 0$, F is anti-parallel to dz, system returns to the equilibrium. This corresponds to stable equilibrium. If, on the other hand $(\partial^2\phi/\partial z^2)|_{z=z_0} < 0$, F is parallel to dz, system is in static equilibrium and possess constant energy. The stability may be investigated by enquiring whether any perturbation decreases the potential energy of the system (which gets fed into kinetic energy). Should there be such perturbation the system is said to be unstable, as it enables to decrease its potential energy which goes into increasing the kinetic energy.

© The Author(s), under exclusive license to Springer Nature Switzerland AG 2022
S. R. Jain et al., *A Primer on Fluid Mechanics with Applications*,
https://doi.org/10.1007/978-3-031-20487-6_11

11.2   Normal Mode Analysis

Linearizing the governing equations, which are of the form:

$$m \frac{d^2 (\delta z)}{dt^2} = - \left(\frac{\partial^2 \phi}{\partial z^2} \right) \bigg|_{z=z_0} dz. \tag{11.2}$$

We seek a solution with time-dependence, $\sim e^{\omega t}$ and then look for a sign of ω. For this purpose, if we express any physical variable q,

$$q = q_0 + q_1(z) \, e^{i(k_x x + k_y y) + \omega t}, \qquad q_1 |z| \ll q_0. \tag{11.3}$$

Here, k_x, k_y are wavenumbers in $x-$ and $y-$ directions; and ω determines the growth rate via the dispersion relation, $\omega(k)$. A real positive (negative) value implies that there is exponential growth (decay); a pure imaginary value implies an oscillatory motion. A combination of these can be inferred for a complex value.

11.3   Jeans Instability

Jeans discovered a simple instance of a gravitational instability which is relevant for formation of cosmic objects. For this purpose Jeans studied an infinite, homogeneous medium with uniform fluid density, $\rho_0 = $ constant at rest ($v_0 = 0$), i.e., no velocity in the basic unperturbed state. For a homogeneous medium, infinite in all directions, the unperturbed pressure, p_0 and density, ρ_0 are independent of position, and likewise the gravitational potential, ϕ_0, satisfying Poisson equation,

$$\nabla^2 \phi = 4\pi \epsilon \rho_0, \qquad (\mathbf{g} = -\nabla \phi) \tag{11.4}$$

The hydrostatic equilibrium,

$$\nabla p_0 = 0 = -\rho \nabla \phi_0 \tag{11.5}$$

implies $\nabla \phi_0 = 0$ for uniform ρ_0. In the present case, the unperturbed state is in static equilibrium ($\mathbf{v}_0 = 0$).

Assuming a small perturbation in density in this infinite medium, the propagation of perturbed velocity $\mathbf{v}_1$ may be calculated using the linearized form of governing equations: Note that in the absence of effects due to gravitational perturbations, the problem would reduce to the classical one of propagation of sound waves with velocity, $c_s = \sqrt{\gamma p_0 / \rho_0}$. However, if the change in gravitational potential consequent to the density perturbation is taken into account, the velocity of propagation will clearly be altered by the effects of gravitation.

The equations of motion are as follows:

$$\rho\frac{dv}{dt} = -\nabla p + \rho\mathbf{g} = -\nabla p - \rho\nabla\phi \quad \text{(momentum conservation)},$$

$$\frac{\partial\rho}{\partial t} + (\mathbf{v}.\nabla)\rho = -\rho\nabla.\mathbf{v} \quad \text{(mass conservation)},$$

$$\nabla^2\phi = 4\pi G\rho \quad \text{(Poisson equation)}$$

$$P = \kappa\rho^\gamma, \quad \text{(Energy (adiabatic) equation)} \tag{11.6}$$

Thus, $(1/p)dp/dt = (\gamma/\rho)d\rho/dt$.

To linearize the system of equations, let us add a correction term to the equilibrium value:

$$p = p_0 + p_1, \quad \rho = \rho_0 + \rho_1, \quad \phi = \phi_0 + \phi_1, \quad \mathbf{v} = \mathbf{v}_1 \tag{11.7}$$

as v_0 is zero in static equilibrium.

$$\frac{d\mathbf{v}}{dt} = \frac{\partial\mathbf{v}}{\partial t} + (\mathbf{v}_1.\nabla)\mathbf{v}_0 + (\mathbf{v}_0.\nabla)\mathbf{v}_1,$$

$$\rho_0\frac{\partial\mathbf{v}_1}{\partial t} + \rho_1\frac{\partial\mathbf{v}_0}{\partial t} = -\nabla p_0 - \nabla p_1 - \rho_0\nabla\phi_1 - \rho_1\nabla\phi_0. \tag{11.8}$$

The second equation becomes, after ignoring second order terms,

$$\rho_0\frac{\partial\mathbf{v}_1}{\partial t} = -\nabla p_1 - \rho_0\nabla\phi_1. \tag{11.9}$$

The other linearized equations are

$$\frac{\partial\rho_1}{\partial t} = -\rho_0\nabla.\mathbf{v}_1,$$

$$\nabla^2\phi_1 = 4\pi G\rho_1,$$

$$p_1 = c_s^2\rho_1, \tag{11.10}$$

where $c_s^2 = \gamma p_0/\rho_0$, and $p_0 = \kappa\rho_0^\gamma$, $p_1 = \kappa\rho_1^\gamma$. Inserting the expression of ρ_1 and taking the divergence of the equation, we get after some manipulation:

$$\frac{\partial^2\rho_1}{\partial t^2} = c_s^2\nabla^2\rho_1 + (4\pi G\rho_0)\rho_1. \tag{11.11}$$

Taking Fourier components by writing $\rho_1 \propto e^{i(\mathbf{k}.\mathbf{r}+\omega t)}$, we get dispersion relation

$$\omega^2 = 4\pi G\rho_0 - c_s^2 k^2. \tag{11.12}$$

Clearly, this leads to an instability provided $\omega^2 > 0$, or, $4\pi G\rho_0 > c_s^2 k^2$, i.e., when the gravitational force dominates the pressure force. For instability,

$$k^2 < k_J^2 = \frac{4\pi G\rho_0}{c_s^2}, \tag{11.13}$$

or, $\Lambda > \lambda_{\text{Jeans}} = 2\pi/k_J$. We obtain the condition that instability requires that the wavelength must exceed Jeans' wavelength, i.e.,

$$\lambda > \lambda_J = \sqrt{\frac{\pi}{G\rho_0}} c_s. \tag{11.14}$$

Thus, the critical Jeans' wavelength for gravitational instability to set in is

$$\lambda_{\text{Jeans}} \sim \sqrt{\frac{T_0}{\rho_0}}, \tag{11.15}$$

note that above a certain density, there is no instability.

We can estimate the Jeans' mass,

$$M_J \sim \rho_0 \lambda_J^3 \propto \rho_0 \left(\frac{T}{\rho_0}\right)^{3/2}. \tag{11.16}$$

In conclusion, the origin of Jeans instability is clearly associated with long wavelength when the gravitational force dominates the internal pressure forces and the system can go to states of lower energy with the liberation of thermal energy.

Interstellar Cloud
Neutral gas is at a temperature, T_0 of about $100\,\text{K}$. The density, $n \sim 10\,\text{cm}^{-3}$, density, $\rho_0 \sim 10^{-23}$ g/cc. The Jeans' wavelength is then

$$\lambda_J \sim 10^{20}\,\text{cm}. \tag{11.17}$$

Jeans' gravitational instability criterion gets altered if we include the contribution of external forces such as rotation and magnetic field in the problem. If we include uniform rotation in the analysis, the presence of Coriolis force in the analysis modifies the dispersion relation to give the solutions of the type, $e^{i(kz+\omega t)}$:

$$\omega^4 + (4\Omega^2 + c_s^2 k^2 - 4\pi G\rho_0)\omega^2 + 4\Omega^2(c_s^2 k^2 - 4\pi G\rho_0)\cos^2\theta = 0 \tag{11.18}$$

where θ is the angle between the direction of Ω and the propagation direction, $\mathbf{k}$ taken to be the z-axis, so that $\Omega = (0, \Omega_y, \Omega_z) = (0, \Omega \sin\theta, \Omega \cos\theta)$. The foregoing quartic equation has the following two roots:

$$\omega^2 = 4\pi G\rho_0 - c_s^2 k^2 - 4\Omega^2 \qquad (11.19)$$

11.3.1 Case of Uniform Rotation

The angular frequency is, where the angle θ is between the vector, Ω and wave vector, $\mathbf{k}$. Propagation of waves in the z-direction is governed by the equation in the reference frame which is rotating with a uniform angular velocity, Ω

$$\rho\frac{d\mathbf{v}}{dt} = -\nabla p - \rho\nabla\phi + 2\rho\mathbf{v} \times \Omega. \qquad (11.20)$$

The other equations along with this are:

$$\frac{d\rho}{dt} + \rho\nabla.\mathbf{v} = 0,$$

$$\nabla^2\phi = 4\pi G\rho, \qquad p = k\rho^\gamma. \qquad (11.21)$$

We linearize the equations by writing

$$p = p_0 + p_1, \quad \rho = \rho_0 + \rho_1, \quad \mathbf{v} = \mathbf{v}_0 + \mathbf{v}_1, \quad \Omega = (0, \Omega_y, \Omega_z). \qquad (11.22)$$

Component-wise, the governing equations are

$$\frac{\partial v_{1x}}{\partial t} - 2v_{1y}\Omega_z + 2\Omega_y v_{1z} = 0,$$

$$\frac{\partial v_{1y}}{\partial t} + 2v_{1x}\Omega_z = 0,$$

$$\frac{\partial v_{1z}}{\partial t} + \frac{c_s^2}{\rho_0}\frac{\partial p_1}{\partial z} + \frac{\partial\phi_1}{\partial z} - 2\Omega_y v_{1x} = 0,$$

$$\frac{\partial\rho_1}{\partial t} + \rho_0\frac{\partial v_{1z}}{\partial z} = 0,$$

$$\frac{\partial^2\phi_1}{\partial z^2} - 4\pi G\rho_1 = 0. \qquad (11.23)$$

We look for solutions of the form, $e^{i(kz+\omega t)}$, leading to the equations for the perturbations:

$$\omega v_{1x} + 2\Omega_y v_{1z} - 2\Omega_z v_{1y} = 0,$$

$$\omega v_{1y} + 2\Omega_z v_{1x} = 0,$$

$$\omega v_{1z} + \frac{c_s^2}{\rho_0} ik\rho_1 + ik\phi_1 - 2\Omega_y v_{1x} = 0,$$

$$\omega \rho_1 + ik\rho_0 v_{1z} = 0,$$

$$-k^2\phi_1 - 4\pi G\rho_1 = 0. \tag{11.24}$$

These equations can be cast in a matrix form where a 5×5 coefficient matrix acts on a column vector, $(v_{1x}, v_{1y}, v_{1z}, \rho_1, \phi_1)^T$. For consistent solutions, the determinant of the coefficient matrix must be zero, ensuing thus the equation for frequencies:

11.4 Rayleigh-Taylor Instability

This is related to the stability of a horizontal interface between two fluids of uniform densities ρ_1 and ρ_2 under gravity. In a uniform gravitational field such an instability occurs when a heavier fluid on the top is supported by a lighter fluid from below. Assume the two fluids to be incompressible separated by an interface that is a plane surface, $z = 0$ which undergoes a small perturbation of the form,

$$\eta(x, t) = Ae^{i(kx+\omega t)}. \tag{11.25}$$

Within the two fluids the motion is supposed to be initially irrotational, so that the velocity u may be expressed as gradient of a potential, ψ and for irrotational flow, curl $u = 0$. The stream functions ψ_1 and ψ_2 for irrotational motion satisfy the equations,

$$\nabla^2\psi_1 = 0, \qquad \nabla^2\psi_2 = 0. \tag{11.26}$$

At the interface, $u_z = (-\partial\psi_1/\partial z)_{z=0} = \partial\eta/\partial t$ in the region 1 and $u_z = (-\partial\psi_2/\partial z)_{z=0} = \partial\eta/\partial t$. Note that $|\nabla\psi_1| \to 0$ as $z \to -\infty$ and $|\nabla\psi_2| \to 0$ as $z \to \infty$ which enables us to write the stream functions in the two regions as,

$$\psi_1 = C_1 \exp[i(kx + kz + \omega t)], \quad z < 0,$$

$$\psi_2 = C_2 \exp[i(kx - kz + \omega t)], \quad z > 0. \tag{11.27}$$

The kinematic boundary condition implies $-c_1 k = A\omega$, $c_2 k = A\omega$. According to the Bernoulli equation, the pressure is given by $(p/\rho) - d\psi/dt + u^2/2 + gz =$ constant in either fluid 1 and 2, neglecting the second order term, $u^2/2$, and retaining in first order,

$$\frac{d\psi}{dt} = \frac{\partial \psi}{\partial t} + (\boldsymbol{u}.\nabla)\psi \simeq \frac{\partial \psi}{\partial t}. \tag{11.28}$$

Now,

$$\left(\frac{P_1}{\rho_1}\right)_{\text{interface}} = \left.\frac{\partial \psi_1}{\partial t}\right|_{z=0} - g\eta,$$

$$\left(\frac{P_2}{\rho_2}\right)_{\text{interface}} = \left.\frac{\partial \psi_2}{\partial t}\right|_{z=0} - g\eta. \tag{11.29}$$

Requiring the pressure balance $p_1 = p_2$ at the interface for all values of z, we get

$$P_1 = -\frac{A\omega^2}{k}\rho_1 - gA\rho_1, \quad P_2 = -\frac{A\omega^2}{k}\rho_2 - gA\rho_2. \tag{11.30}$$

This gives the dispersion relation,

$$\omega^2 = gk\frac{\rho_2 - \rho_1}{\rho_2 + \rho_1}. \tag{11.31}$$

Clearly, $\omega^2 > 0$ for the system to be unstable if $g(\rho_2 - \rho_1) > 0$, or, $\rho_2 > \rho_1$, that is, the fluid layers are basically unstable if the heavier fluid, ρ_2, rests on the lighter fluid, ρ_1, resulting in faster instability for larger k (or, smaller wavelength).

The hydrodynamic analysis was extended by the Schwarzschild and Harm to discuss the hydromagnetic Rayleigh-Taylor instability for two incompressible, inviscid fluids of uniform densities ρ_1 and ρ_2 with the heavier fluid of density ρ_2 resting on top of the lighter fluid of density ρ_1 separated by a horizontal boundary under force of gravity, acting vertically downwards and carrying a uniform magnetic field, $B_0\hat{x}$ underneath in the lighter fluid.

Supplementing the governing hydrodynamic equations by Maxwell equations, the dispersion relation becomes

$$\omega^2 = gk\frac{\rho_2 - \rho_1}{\rho_2 + \rho_1} - \frac{B_0^2 k_x^2}{2\pi(\rho_2 + \rho_1)}. \tag{11.32}$$

This shows that the magnetic field produces no additional effect when the wavenumber, $\mathbf{k}$ is normal to the magnetic field, i.e., $k_x = 0$. However, ripples along the field, $k = k_x$ produces a restoring force through the magnetic tension which permits instability, i.e. $\omega^2 > 0$ only when $k < k_{\text{critical}} = 2\pi g(\rho_2 - \rho_1)/B_0^2$. Note when

$\rho_2 > \rho_1$ we get convective instability and for $\rho_2 < \rho_1$, we have the oscillatory, surface modes.

11.5   Kelvin-Helmholtz Instability

The classical Kelvin-Helmholtz instability arises when an inviscid flow slides over the surface of a heavier fluid at rest. When the interface between two fluids is slightly perturbed, the resulting dynamical forces may be able to overcome the stabilizing effect due to gravity leading to surface waves. Let ρ_1 and ρ_2 be the respective uniform densities of the lower and upper fluids with $\rho_2 > \rho_1$ and u the relative streaming velocity. Choose axes moving with the mean velocity of the two streaming fluids.

Introducing the velocity potential, ψ, we know that curl $\boldsymbol{u} = 0$ for irrotational motion and using the mass conservation, div $\boldsymbol{u} = 0$ gives $\nabla^2 \psi = 0$.

Suppose the interface $z = 0$, in the absence of surface tension, is perturbed to have the form, $\eta = \eta(x, t)$. Then, the velocity potential, $\boldsymbol{u} = -\nabla \psi$, $\psi = U/2 + \psi_1$ $(z < 0)$, $\psi = -U/2 + \psi_2$, $(z > 0)$.

Assume the Fourier decomposition for ψ and η of the form, $e^{i(kx+\omega t)}$. Noting that within the two fluids, the motion is irrotational, so that $\nabla^2 \psi_1 = 0$ and $\nabla^2 \psi_2 = 0$, implying the velocity potentials to have the form

$$\psi_1 = C_1 \exp[ikx + kz + \omega t], \quad z < 0,$$

$$\psi_2 = C_2 \exp[ikx - kz + \omega t], \quad z > 0. \tag{11.33}$$

If the z-displacement of the free surface is η, then the kinematic boundary condition gives

$$u_z = \frac{d\eta}{dt} = \frac{\partial \eta}{\partial t} + (\boldsymbol{u_0}.\nabla)\eta, \text{ or}$$

$$u_{z2} = -\frac{\partial \psi_2}{\partial z}\bigg|_{z=0} = \frac{\partial \eta}{\partial t} + (\boldsymbol{u}.\nabla)\eta, \text{ and },$$

$$\text{or, } -kC_1 = \left(\omega - \frac{1}{2}ikU\right)A, \ z < 0,$$

$$kC_2 = \left(\omega + \frac{1}{2}ikU\right)A, \ z > 0. \tag{11.34}$$

From Bernoulli's equation, $P/\rho - d\psi/dt + u^2/2 + gz = $ constant, we get to first order,

$$\left(\frac{P}{\rho_1}\right)_{z=0} = \left(\frac{d\psi_1}{dt}\right)_{z=0} - g\eta,$$

$$\left(\frac{P}{\rho_2}\right)_{z=0} = \left(\frac{d\psi_2}{dt}\right)_{z=0} - g\eta. \tag{11.35}$$

The continuity of pressure across the interface, $z = 0$ gives

$$\rho_1\left(\omega - \frac{1}{2}ikU\right)C_1 = \rho_2\left(\omega + \frac{1}{2}ikU\right)C_2 + g(\rho_1 - \rho_2)A. \tag{11.36}$$

Eliminating C_1 and C_2 between Eqs. (11.34) and (11.36), we recover the dispersion relation,

$$\rho_1\left(\omega - \frac{1}{2}ikU\right)^2 + \rho_2\left(\omega + \frac{1}{2}ikU\right)^2 + gk(\rho_1 - \rho_2) = 0,$$

or, $\omega^2(\rho_1 + \rho_2) - i\omega kU(\rho_1 - \rho_2) + gk(\rho_1 - \rho_2) - \dfrac{(\rho_1 + \rho_2)}{4}k^2U^2 = 0.$

$$\tag{11.37}$$

This gives a solution,

$$\frac{\omega}{kU} = \frac{i}{2}\left(\frac{\rho_1 - \rho_2}{\rho_1 + \rho_2}\right) \pm \left\{\frac{1}{4} - \frac{1}{4}\frac{\rho_2 - \rho_1}{\rho_1 + \rho_2} - \frac{g}{kU^2}\frac{\rho_2 - \rho_1}{\rho_1 + \rho_2}\right\}. \tag{11.38}$$

The instability criterion then becomes

$$U^2 > \frac{\rho_1^2 - \rho_2^2}{\rho_1\rho_2}\frac{g}{k}. \tag{11.39}$$

Note that in the absence of any streaming motion, $U = 0$, we get

$$\omega^2 = gk\frac{\rho_2 - \rho_1}{\rho_1 + \rho_2} \tag{11.40}$$

which is the dispersion relation for the Rayleigh-Taylor instability.

A classic example is the waves on the ocean surface separating air and water masses, giving the streaming velocity, $U \sim 650$ cm/s. There are several instances in geophysical and astrophysical fluids of different densities co-existing in relative motion, e.g. winds, blowing over the ocean surface, cometary tails being swept by the solar wind, accretion flow past compact objects.

11.6   Rayleigh-Bènard Convection

We summarize here a formulation of the stability of a fluid layer heated from below under the gravitational force in the presence of adverse temperature gradient [19]. For this purpose, we consider the following set of governing equations in the Boussinesq approximation,

$$\text{(momentum)} \quad \frac{d\boldsymbol{u}}{dt} = -\frac{\nabla p}{\rho_0} - \left(1 + \frac{\delta\rho}{\rho_0}g\hat{z} + \nu\nabla^2\boldsymbol{u}\right),$$

$$\text{(mass)} \quad \nabla.\boldsymbol{u} = 0, \tag{11.41}$$

ν being the kinematic viscosity, and assuming variation in the density ρ only as a result of variations in the temperature and not pressure, so that

$$\delta\rho = \left(\frac{\partial\rho}{\partial T}\right)_P \delta T = -\rho_0\alpha\delta T \tag{11.42}$$

where α is the coefficient of volume expansion. Finally, the energy equation taken in the form,

$$\text{energy} \quad \rho_0 C_V \frac{dT}{dt} + p\nabla.\boldsymbol{u} = \nabla.K\nabla T, \tag{11.43}$$

K being the coefficient of thermal conductivity and C_V, the specific heat at constant volume. Adopting the unperturbed state of the physical variables, $T_0(z) = T_0 - \beta z$, β being the constant temperature gradient, $\rho_0(z) = \rho_0(1 + \alpha\beta z)$ and $P_0(z) = P_0 - g\rho_0(z + \alpha\beta z^2/2)$.

Linearize the governing equations, using the Boussinesq approximation, we obtain the following set of equations in the first order, denoting the physical variables by subscript U,

$$\frac{d\boldsymbol{u}}{dt} = -\frac{1}{\rho_0}\nabla p_1 + g\alpha T_1\hat{z} + \nu\nabla^2\boldsymbol{u}_1,$$

$$\nabla.\boldsymbol{v} = 0,$$

$$\frac{\partial T_1}{\partial t} = U_{1z}\beta + \kappa\nabla^2 T_1, \tag{11.44}$$

κ being thermal diffusivity, $K/\rho_0 C_V$, U_{1z} is the vertical velocity component. Take curl(curl) of the equation of motion to obtain

$$\frac{\partial}{\partial t}\nabla \times \nabla \times \boldsymbol{u} = \nabla \times \nabla \times (g\alpha T_1 \hat{z}), \text{ or,}$$

$$\frac{\partial}{\partial t}(-\nabla^2 \boldsymbol{u}) = \nabla \times \nabla \times (g\alpha T_1 \hat{z}). \tag{11.45}$$

Again, differentiating w.r.t. t we recover the equation for z-component of the velocity U_{1z} in the form

$$\frac{\partial}{\partial t^2}(\nabla^2 U_{1z}) = g\alpha \left(\frac{\partial^2}{\partial x^2} + \frac{\partial^2}{\partial y^2}\right)\frac{\partial T_1}{\partial t},$$

$$\text{and } \frac{\partial T_1}{\partial t} = U_{1z} + \kappa \nabla^2 T_1. \tag{11.46}$$

Fourier analyzing the variables U_{1z}, T_1 by writing

$$U_{1z} = U_1(z)e^{(ik_x x + ik_y y) + \omega t},$$

$$T_1 = T_1(z)e^{(ik_x x + ik_y y) + \omega t}, \tag{11.47}$$

ω being the growth rate and adopting the rigid boundary conditions, U_{1z} and $T_1 = 0$ at $z = 0$ and $z = d$, holding the boundaries at constant temperature, and assuming no motion across them, by taking

$$U_{1z} = A \sin\frac{n\pi z}{d}e^{(ik_x x + ik_y y) + \omega t},$$

$$T_1 = B \sin\frac{n\pi z}{d}e^{(ik_x x + ik_y y) + \omega t}, \tag{11.48}$$

and inserting these in the linearized equations we obtain the dispersion relation,

$$\omega^2 + (\kappa + \nu)k^2\omega - g\alpha\beta\frac{k_H^2}{k^2} = 0, \tag{11.49}$$

where $k_H^2 = k_x^2 + k_y^2$ and $k^2 = k_H^2 + n^2\pi^2/d^2$. This gives, after maximizing the growth rate, ω, w.r.t. the horizontal wavenumber k_H, which corresponds to the fastest growing mode. In the absence of kinematic viscosity ν, but in the presence of thermal diffusivity, κ, however, the maximal growth rate turns out to be $\omega_{max} = g\alpha\beta(n^2\pi^2/d^2)/k\nu$ corresponding to $k_H^{max} = \kappa k^3\omega/(g\alpha\beta)$.

It is useful to introduce in these formalisms the dimensionless Rayleigh number $R = g\alpha\beta d^4/k^2\pi^4$ or $R = g\alpha\beta d^4/\kappa\nu\pi^4$ in the presence of viscosity and thermal diffusivity, which is a measure of the destabilizing buoyancy force.

Suppose we now introduce a uniform magnetic field in the analysis. The linearized momentum equations then become

$$\frac{\partial \boldsymbol{u}}{\partial t} = -\frac{\nabla p_1}{\rho_0} + \frac{\boldsymbol{j}_1 \times \boldsymbol{B}_0}{4\pi} - \frac{\delta\rho}{\rho_0}g\hat{z}, \text{ noting } \boldsymbol{j}_0 = 0,$$

$$\nabla . \boldsymbol{u} = 0,$$

$$\frac{\partial T_1}{\partial t} = U_{1z}\beta + \kappa\nabla^2 T_1, \tag{11.50}$$

where κ is thermal diffusivity.

Adopting the Fourier decomposition in the form, (11.48), the dispersion relation for uniform vertical magnetic field $B_0\hat{z}$, becomes

$$\omega'^3 + \omega'^2(\alpha_H^2 + 1) + \omega'\left\{Q - R\frac{\alpha_H^2}{\alpha_H^2 + 1}\right\} + Q(\alpha_H^2 + 1) = 0, \tag{11.51}$$

where $\alpha_H^2 = (k_x^2 + k_y^2)d^2/\pi^2$ is the dimensionless horizontal wavenumber and the growth rate, ω' is expressed in units of $\omega_0 = \pi^2\kappa^2/d^2$, $Q = B_0^2 d^2/(4\pi^3\rho_0\kappa^2)$ is the Chandrasekhar number which is a measure of magnetic restoring force against dissipation, and R, the dimensionless Rayleigh number which measures the destabilizing buoyancy force against dissipation.

Likewise for a uniform horizontal magnetic field, $B_0\hat{x}$, the dispersion relation is

$$\omega^3 + \omega^2(\alpha_X^2 + 1) + \omega\left\{Q\frac{\alpha_X^2}{2} - R\frac{\alpha_H^2}{\alpha_H^2 + 1}\right\} + Q\frac{\alpha_X^2}{2}(\alpha_X^2 + 1) = 0. \tag{11.52}$$

These cubic equations for ω have three roots, one of which is always real, positive, and the other two form a complex conjugate pair. A typical plot of the roots in the $R - Q$ plane is shown in Fig. 11.1.

11.7 Parker Instability and Magnetic Buoyancy

The interstellar and intergalactic media are known to be clumpy and non-uniformly distributed, made up of different phases: cold, warm, hot in equilibrium with one another. Parker argued [87] that a uniform distribution of the interstellar medium would be unstable, which is known as Parker instability related to the phenomenon of magnetic buoyancy which is presumably the reason behind the interstellar matter fragmenting into clumps.

Assume the magnetic field of a galaxy to be frozen in the interstellar medium and is uniformly distributed in a layer in the horizontal direction. Suppose now the system undergoes a small perturbation with parts of the magnetic field lines bulging upwards.

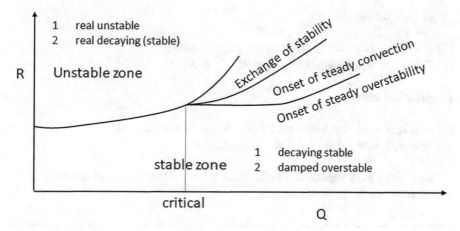

Fig. 11.1 The three roots of the angular frequency are shown here in a plot where Chandrasekhar number, R is plotted against Rayleigh number, Q. There is a stable and an unstable zone in addition to regions of different stabilities

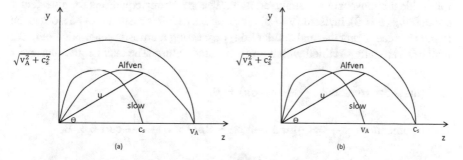

Fig. 11.2 Figure (**a**) corresponds to $v_A > c_s$ and (**b**) to $v_A < c_s$

By symmetry the gravitational field is directed towards the central plane of the layer. Taking the magnetic field to be frozen in the layer, the plasma blobs will slide down vertically in the bulged region and as a result of the slide down from the region of the bulge, this elevated region becomes lighter and more buoyant, and continues to keep rising, We then expect this region to rise up further and the initial bulge keeps getting larger leading to an instability known as Parker's magnetic buoyancy instability. The interstellar clouds get accumulated at either end of the horizontal magnetic flux-tube, and as the magnetic field lines become more bent, the magnetic tension gets increasingly stronger to eventually halt the rise of the upper portion of the bulge [80, 87]. In the context of galaxies, the magnetic field lines rise out of the plane, whereas the interstellar plasma blobs collect in the valleys of the magnetic field lines giving rise to an intermittent clumpy medium (Fig. 11.2).

Transverse Waves

(1) $\theta = 0, u = v_A \implies$ Alfven waves in z-direction,
(2) $\theta = \pi, u \to 0$ (slow), $u \to (v_A^2 + c_S^2)^{1/2}$ (fast).

Longitudinal Waves

(1) $\theta = 0, u = c_S$, ordinary sound wave motion is parallel to $\mathbf{B}_0$
(2) $\theta = \pi/2, u = (v_A^2 + c_S^2)^{1/2}$ (fast wave).

Two of the most significant applications of magnetosonic waves are the Earth's atmosphere and the solar wind.

11.8 Various Wave Modes for an Ideal Atmosphere

Various wave modes in presence of pressure, gravity, rotation and magnetic field for an ideal atmosphere are analyzed here. The governing equations for a perfectly conducting ($\eta = 0$), inviscid ($\nu = 0$) atmosphere which is assumed to be isothermal in the absence of any thermal conduction, possessing a uniform magnetic field, B_0, which is vertically stratified with a uniform temperature. The usual equations are:

$$(\text{mass conservation}) \quad \frac{\partial \rho}{\partial t} + \text{div}(\rho \mathbf{u}) = 0,$$

$$(\text{momentum}) \quad \rho \frac{d\mathbf{u}}{dt} = -\text{grad } p + \rho \mathbf{g} - 2\rho \mathbf{\Omega} \times \mathbf{u} + \frac{1}{4\pi \rho}\text{curl } \mathbf{B} \times \mathbf{B},$$

$$(\text{energy}) \quad \frac{d}{dt}\left(\frac{p}{\rho^\gamma}\right) = 0, \quad \text{or}, \quad \frac{1}{p}\frac{dp}{dt} = \frac{\gamma}{\rho}\frac{d\rho}{dt},$$

$$(\text{induction equation}) \quad \frac{\partial \mathbf{B}}{\partial t} = \text{curl } (\mathbf{u} \times \mathbf{B}), \quad \text{div } \mathbf{B} = 0, \quad \text{curl } \mathbf{H} = 4\pi \mathbf{J}/c,$$

$$(11.53)$$

neglecting the displacement current and assuming $\gamma = C_p/C_v$ to be constant. Consider an equilibrium state: p_0, ρ_0, T_0, B_0 with $\mathbf{u}_0 = 0$ and a small perturbation from equilibrium,

$$p = p_0 + p_1, \quad \rho = \rho_0 + \rho_1, \quad \mathbf{B} = \mathbf{B}_0 + \mathbf{b}_1, \quad \mathbf{u} = \mathbf{u}_0 + \mathbf{u}_1. \quad (11.54)$$

The linearized equations then become

$$\frac{\partial \rho_1}{\partial t} + \mathbf{u}_1.\text{grad}\rho_0 + \rho_0\text{div}\mathbf{u}_1 = 0,$$

$$\rho_0 \frac{d\mathbf{u}_1}{dt} = -\text{grad } p_1 + \rho_1 \mathbf{g} - 2\rho_0 \mathbf{\Omega} \times \mathbf{u}_1 + \frac{1}{4\pi\rho_0}\text{curl } \mathbf{b}_1 \times \mathbf{B}_0,$$

$$\frac{dp_1}{dt} + \mathbf{u}_1.\text{grad}p_1 = c_s^2 \left(\frac{d\rho_1}{dt} + \mathbf{u}_1.\text{grad}\rho_0 \right) = 0,$$

$$\frac{\partial \mathbf{b}_1}{\partial t} = \text{curl}\,(\mathbf{u}_1 \times \mathbf{B}_0), \quad \text{div}\,\mathbf{b}_1 = 0. \tag{11.55}$$

This set of equations may be reduced to a single differential equation for the variable, $\mathbf{u}_1$ by differentiating momentum equation w.r.t. t and substituting $\partial p_1/\partial t$, $\partial \rho_1/\partial t$, and $\partial \mathbf{b}_1/\partial t$ after using the equilibrium profiles for $\rho_0(z)$, $P_0(z)$, to get [93]

$$\frac{\partial^2 \mathbf{u}_1}{\partial t^2} = c_s^2 \nabla (\text{div}\,\mathbf{u}_1) - (\gamma - 1)g\hat{z}(\text{div}\,\mathbf{u}_1) - g\nabla u_{1z}$$

$$- 2\Omega \times \frac{\partial \mathbf{u}_1}{\partial t} + [\text{curl}\,(\text{curl}\,(\mathbf{u}_1 \times \mathbf{B}_0))] \times \frac{\mathbf{B}_0}{4\pi \rho_0}. \tag{11.56}$$

Introducing the Fourier decomposition for perturbation in velocity, $u_1 e^{i(\mathbf{k}.\mathbf{r} - \omega t)}$ where $\mathbf{k}$ is the wave vector and ω the frequency by replacing $\partial/\partial t$ by $-i\omega$ and ∇ by ik to get finally the dispersion relation:

$$\omega^2 \mathbf{u}_1 = c_s^2 \mathbf{k}(\mathbf{k}.\mathbf{u}_1) + i(\gamma - 1)g\hat{z}(\mathbf{k}.\mathbf{u}_1) + igu_{1z}\mathbf{k} - 2i\omega\Omega \times \mathbf{u}_1$$

$$+ \mathbf{k} \times [(\mathbf{k} \times (\mathbf{u}_1 \times \mathbf{B}_0))] \times \frac{\mathbf{B}_0}{4\pi \rho_0}. \tag{11.57}$$

Assume the wavelength of perturbation, $2\pi/k$ is small compared to the local scale-height, $H_p = c_s^2/\gamma g$ where ρ_0 may be regarded as locally constant adopting the WKBJ approximation ($kH_p \gg 1$), we analyze the following special cases for different wave modes:

1. *Acoustic modes*
   With $\mathbf{B} = 0$, $\mathbf{g} = 0$, $\Omega = 0$, the only restoring force being the pressure gradient, the dispersion relation reduces to $\omega^2 \mathbf{u}_1 = c_s^2(\mathbf{k}.\mathbf{u}_1)\mathbf{k}$. Trivially, $\mathbf{k}.\mathbf{u}_1 = 0$ gives $\omega^2 = 0$. But for $\mathbf{k}.\mathbf{u}_1 \neq 0$, taking scalar product with $\mathbf{k}$, we get $\omega^2 = k^2 c_s$ yielding acoustic waves owing their existence to compressibility, which propagate equally in all directions at a phase speed $\mathbf{v}_p = \omega/k = c_s$ and the group velocity, $\mathbf{v}_s = d\omega/dk = c_s$ in the direction of $\mathbf{k}$.
2. *Hydromagnetic modes*
   With $\Omega = 0$, $\mathbf{g} = 0$, p (or c_s^2) $= 0$, when magnetic forces dominate the equilibrium to give the dispersion relation,

$$\omega^2 \mathbf{u}_1 = \mathbf{k} \times [(\mathbf{k} \times (\mathbf{u}_1 \times \mathbf{B}_0))] \times \hat{\mathbf{B}}_0 \, v_A, \tag{11.58}$$

$\hat{\mathbf{B}}_0$ being the unit vector in the direction of $\mathbf{B}_0$ and $v_A^2 = B_0^2/4\pi\rho_0$, or,

$$\frac{\omega^2}{v_A^2}\mathbf{u}_1 = (\mathbf{k}.\hat{\mathbf{B}}_0)^2\mathbf{u}_1 - (\mathbf{k}.\mathbf{u}_1)(\mathbf{k}.\hat{\mathbf{B}}_0)\hat{\mathbf{B}}_0 + [(\mathbf{k}.\mathbf{u}_1) - (\mathbf{k}.\hat{\mathbf{B}}_0)(\hat{\mathbf{B}}_0.\mathbf{u}_1)]\mathbf{k}.$$

(11.59)

Denote by θ_B the angle between $\mathbf{k}$ and $\mathbf{B}_0$ to give

$$\frac{\omega^2}{v_A^2}\mathbf{u}_1 = k^2\cos^2\theta_B\mathbf{u}_1 - (\mathbf{k}.\mathbf{u}_1)k\cos\theta_B\hat{\mathbf{B}}_0 + [(\mathbf{k}.\mathbf{u}_1) - k\cos\theta_B\,(\hat{\mathbf{B}}_0.\mathbf{u}_1)]\mathbf{k}.$$

(11.60)

Noting that div $\mathbf{b}_1 = 0$ implies that $\mathbf{k}.\mathbf{b}_1 = 0$, so that magnetic field perturbation is normal to the direction of propagation.

Next, the scalar product with $\hat{\mathbf{B}}_0$ yields

$$\hat{\mathbf{B}}_0.\mathbf{u}_1 = 0,$$

(11.61)

since the RHS vanishes, since $\mathbf{k}.\hat{\mathbf{B}}_0 = k\cos\theta_B$. Finally, the scalar product with $\mathbf{k}$ gives

$$(\omega^2 - k^2v_A^2)(\mathbf{k}.\mathbf{u}_1) = 0$$

(11.62)

which has two distinct solutions:

1. $\mathbf{k}.\mathbf{u}_1 = 0$ for an incompressible perturbation, div $\mathbf{u}_1 = 0$ reduces the dispersion relation to $\omega^2 = k^2v_A^2\cos^2\theta_B$, or $\omega = \pm kv_A\cos\theta_B$ for *shear* Alfvén waves with phase speed $v_A\cos\theta_B$; while $\mathbf{k}.\mathbf{u}_1 = 0$ implies that Alfvén waves are transverse in the sense that the velocity perturbation is normal to the propagation direction.
2. The second solution, viz., $\omega^2 = k^2v_A^2$ for $\mathbf{k}.\mathbf{u}_1 \neq 0$, results in compressional Alfvén waves with phase velocity, v_A, regardless of the propagation angle. In the special circumstance of the propagation along the magnetic field, i.e., $\theta_B = 0$, we recover transverse compressional Alfvén waves, identical to the ordinary Alfvén wave driven wholly by magnetic tension.
3. *Internal gravity waves*

   For the case of $\Omega = 0$ and $\mathbf{B}$ from (11.56), taking the scalar product with $\mathbf{k}$ and $\hat{z}$ in turn and gathering terms in u_{1z} and $\mathbf{k}.\mathbf{u}$, we obtain

$$igk^2u_{1z} = (\omega^2 - k^2c_s^2 + i(\gamma - 1)gk_z)(\mathbf{k}.\mathbf{u}_1), \text{ and}$$

$$(\omega^2 - igk_z)u_{1z} = (k^2c_s^2 + i(\gamma - 1)gk_z)(\mathbf{k}.\mathbf{u}_1).$$

(11.63)

Eliminating $(\mathbf{k}.\mathbf{u}_1)/u_{1z}$ between these two equations, the resulting dispersion relation becomes

$$\omega^4 - \omega^2 k^2 c_s^2 + g^2(\gamma - 1)k_z^2 + \omega^2 i(\gamma - 1)gz = igk^2 c_s^2 k_z - g^2 k^2(\gamma - 1). \tag{11.64}$$

The objective is to seek wave solutions with frequencies of order of Brunt-Väisälä frequency N_{B-V} and much slower than that of sound waves, so that $\omega = g/c_s \ll kc_s$. This essentially implies the wavelength involved is much smaller than a typical scale-height and the dispersion equation reduces to $\omega^2 c_s^2 \simeq (\gamma - 1)g^2(1 - k_z^2/k^2)$.

In terms of N_{B-V} and the inclination, $\theta_g = \cos^{-1}(k_z/k)$ between the propagation direction and the z-axis, this may be written as $\omega = N \sin \theta_g$ for "internal" gravity waves; surface gravity waves propagate along the surface between two fluids.

4. *Inertial waves due to Coriolis force*

For $\mathbf{g} = 0$, $\mathbf{B}_0 = 0$, the basic equation reduces to

$$\omega^2 \mathbf{u}_1 = c_s^2 k^2 (\mathbf{k}.\mathbf{u}_1)\hat{\mathbf{u}}_1 - 2i\omega\Omega \times \mathbf{u}_1. \tag{11.65}$$

The incompressible solution, $\mathbf{k}.\mathbf{u}_1 = 0$ gives

$$\omega^2 \mathbf{u}_1 = -2i\omega\Omega \times \mathbf{u}_1. \tag{11.66}$$

and the vector product with $\mathbf{k}$ then yields

$$\omega^2 \mathbf{k} \times \mathbf{u}_1 = 2i\omega(\mathbf{k}.\Omega)\mathbf{u}_1. \tag{11.67}$$

Taking the cross product with $\mathbf{k}$ implies

$$\omega^2 \mathbf{k} \times (\mathbf{k} \times \mathbf{u}_1) = 2i\omega(\mathbf{k}.\Omega)(\mathbf{k} \times \mathbf{u}_1)$$
$$= -4(\mathbf{k}.\Omega)^2|\mathbf{u}_1|^2, \tag{11.68}$$

or, $\omega = \pm(\mathbf{k}.\Omega)$ as the dispersion relation or inertial waves which may be expressed in terms of the angle θ_Ω between the rotation axis and propagation direction, $\omega = \pm 2\Omega \cos \theta_\Omega$, thus implying that as the wave propagated, the velocity vector rotates about the direction of propagation. These waves are transverse in nature and are circularly polarized. Furthermore, the vorticity $\boldsymbol{\omega} = \operatorname{curl} \mathbf{u}_1$ is either parallel or anti-parallel to the velocity $\mathbf{u}_1$, so that $\mathbf{u}_1.\operatorname{curl} \mathbf{u}_1$ produces the helicity of the flow field.

5. *Effect of the Coriolis force on Alfvén waves in the incompressible fluids*

Note with $\mathbf{g} = 0$, $\mathbf{k}.\mathbf{u}_1 = 0$, the dispersion relation in the case of magnetorotational mode becomes

$$\omega^2\mathbf{u}_1 = -2i\omega\mathbf{\Omega} \times \mathbf{u}_1 + [\mathbf{k} \times (\mathbf{k} \times (\mathbf{u}_1 \times \hat{\mathbf{B}}_0))] \times \hat{\mathbf{B}}_0 v_A^2. \tag{11.69}$$

The vector product of this equation by $\mathbf{k}$ results in the dispersion relation,

$$\omega^2 \mp \omega_I\omega - \omega_A^2 = 0, \tag{11.70}$$

where $\omega_I = 2(\mathbf{k}.\mathbf{\Omega})$ and $\omega_A = \mathbf{k}.\hat{\mathbf{B}}_0\, v_A$ are pure inertial and Alfvén frequencies respectively.

The solution of this equation in the limit when $\omega_I \ll \omega_A$

$$\omega^2 = \omega_A^2 \left(1 \pm \frac{\omega_I}{\omega_A}\right) \tag{11.71}$$

so that the Coriolis force produces a small splitting of Alfvén wave frequency. In the opposite limit, when $\omega_I \gg \omega_A$, the solution becomes $\omega^2 \simeq \omega_I^2$ (inertial wave), $\omega^2 \simeq \omega_A^4/\omega_I^2$ (hydromagnetic inertial wave).

Shock Waves

12

12.1 Conservation Laws and Rankine-Hugoniot Relations

Surfaces across which the velocities and thermodynamic variables (pressure, density, temperature, entropy) jump are called shocks. When the speed u of the fluid is greater than the local speed, c_s, there are two regions of interest.

1. *H I region:* This corresponds to the ratio of specific heats, $\gamma = 5/3$, $T \sim 100$ K, $\mu = 1.5$. The local sound speed is thus

$$c_s \sim \sqrt{\frac{\gamma RT}{\mu}} \sim 1.2\ \text{km/s.} \tag{12.1}$$

2. *H II region:* This corresponds to the ratio of specific heats, $\gamma = 5/3$, $T \sim 10^4$ K, $\mu = 0.7$. The local sound speed is thus

$$c_s \sim \sqrt{\frac{\gamma RT}{\mu}} = \sqrt{\frac{\gamma P}{\rho}} \sim 14\ \text{km/s.} \tag{12.2}$$

Mean dispersion velocity, $\langle v^2 \rangle^{1/2} \sim 7\text{--}8$ km/s. In H I region, mean velocities considerably exceed the local sound speed which implies the occurrence of shock waves.

© The Author(s), under exclusive license to Springer Nature Switzerland AG 2022
S. R. Jain et al., *A Primer on Fluid Mechanics with Applications*,
https://doi.org/10.1007/978-3-031-20487-6_12

For a perfect gas, $P = (R/\mu)\rho T$ where $R/\mu = c_p - c_v$, and $c_s^2 = \gamma P/\rho = \gamma RT/\mu$. The energy

$$E = c_v T = -\{(c_p - c_v) - c_p\} = \left\{\frac{c_p}{c_v} - \frac{c_p - c_v}{c_v}\right\} c_v T$$

$$= \{\gamma - (\gamma - 1)\} c_v T. \tag{12.3}$$

The sound speed is thus

$$c_s^2 = \frac{\gamma RT}{\mu} = \frac{\gamma(c_p - c_v)}{c_v} c_v T = \gamma(\gamma - 1) c_v T. \tag{12.4}$$

Internal energy, $E = c_v T = c_s^2/\gamma(\gamma - 1)$. The enthalpy,

$$I = E + P/\rho = c_p T = c_s^2/(\gamma - 1). \tag{12.5}$$

Let us examine the gas flow field governed by Bernoulli equation, which can be recast as

$$\frac{u^2}{2} + I = \text{constant, or}$$

$$\frac{u^2}{2} + \frac{c_s^2}{\gamma - 1} = \text{constant} = \frac{u_{\max}^2}{2}. \tag{12.6}$$

The $u_{\max}$ is the maximum possible velocity when p (and hence c_s^2) vanishing while c_s^* is the critical sonic speed at which $u = c_s$ and c_{s0} is reached when $u = 0$.

We can cast Bernoulli's equation in the form

$$u^2 - c_s^{*2} = \frac{2}{\gamma + 1}(u^2 - c_s^2) \tag{12.7}$$

which implies that $u > c_s$ ($u < c_s$) according as $u > c_s^*$ ($u < c_s^*$), or, the Mach number, $M = u/c_s$ is > 1, (supersonic) (< 1 (subsonic)).

Consider a one-dimensional flow with streamtubes of unit cross-sectional area across an interface between two regions 1 (rarer) and 2 (denser), with velocities u_1 and u_2, and, other variables (p_1, ρ_1, c_{s1}) and (p_2, ρ_2, c_{s2}). The following conservation laws hold:

1. Mass conservation: $\rho_1 u_1 = \rho_2 u_2 = m$ say.
2. Momentum conservation, where we consider the impulse,

$$(p_1 - p_2)\delta t = (\rho_2 u_2)u_2 \delta t - (\rho_1 u_1)u_1 \delta t \tag{12.8}$$

or

$$p_1 + \rho_1 u_1^2 = p_2 + \rho_2 u_2^2.$$
(12.9)

3. Energy conservation

$$\frac{p_1}{\rho_1} + E_1 + \frac{u_1^2}{2} = \frac{p_2}{\rho_2} + E_2 + \frac{u_2^2}{2}$$
(12.10)

or

$$I_1 + \frac{u_1^2}{2} = I_2 + \frac{u_2^2}{2}.$$
(12.11)

That is,

$$\frac{u_1^2}{2} + \frac{c_s^2}{\gamma - 1} = \frac{\gamma - 1}{2(\gamma + 1)} c_s^{*2}, \text{ or}$$

$$\frac{u_1^2}{2} + \frac{\gamma_1}{\gamma_1 - 1} \frac{p_1}{\rho_1} = \frac{u_2^2}{2} + \frac{\gamma_2}{\gamma_2 - 1} \frac{p_2}{\rho_2}.$$
(12.12)

This set of conservation equations is known as Rankine-Hugoniot relations [63] (where it is customary to take $\gamma_1 = \gamma_2 = \gamma$) from which it follows, noting

$$u_2 - u_1 = \frac{p_1}{\rho_1 u_1} - \frac{p_2}{\rho_2 u_2}$$
(12.13)

to give

$$c_s^2 = \frac{(\gamma + 1)c_s^{*2}}{[2 + M^2(\gamma - 1)]}.$$
(12.14)

From momentum conservation, we deduce

$$\frac{u_1^2}{2} + \frac{\gamma}{\gamma - 1} \frac{p_1}{\rho_1} = \frac{u_2^2}{2} + \frac{\gamma}{\gamma - 1} \frac{p_2}{\rho_2} = \frac{\gamma + 1}{2(\gamma - 1)} c_s^{*2},$$
(12.15)

which gives, after dividing by u_1 and u_2 respectively:

$$\frac{p_1}{\rho_1 u_1} = \frac{(\gamma + 1)}{2\gamma} \frac{c_s^{*2}}{u_1} - \frac{\gamma - 1}{\gamma} \frac{u_1}{2}$$
(12.16)

and

$$\frac{p_2}{\rho_2 u_2} = \frac{(\gamma + 1)}{2\gamma} \frac{c_s^{*2}}{u_2} - \frac{\gamma - 1}{\gamma} \frac{u_2}{2}. \tag{12.17}$$

After subtraction we recover

$$\frac{p_1}{\rho_1 u_1} - \frac{p_2}{\rho_2 u_2} = c_s^{*2} \frac{(\gamma + 1)}{2\gamma} \frac{u_2 - u_1}{u_1 u_2} - \frac{\gamma - 1}{2\gamma}(u_1 - u_2), \tag{12.18}$$

consequently,

$$u_2 - u_1 = c_s^{*2} \frac{(\gamma - 1)}{2\gamma} \frac{u_2 - u_1}{u_1 u_2} + \frac{\gamma - 1}{2\gamma}(u_2 - u_1). \tag{12.19}$$

Hence,

$$(u_1 - u_2) \frac{(\gamma + 1)}{2\gamma} \left\{ \frac{c_s^{*2}}{u_1 u_2} - 1 \right\} = 0, \tag{12.20}$$

which gives $u_1 u_2 = c_s^{*2}$. Note that c_s^{*2} has a constant magnitude in both the regions, 1 and 2 and consequently, either $u_1 > c_s^*$ and $u_2 < c_s^*$, or, $u_1 > c_s^*$ and $u_2 < c_s^*$. That is, the occurrence of a shock is *impossible* unless the flow on one of the sides is supersonic.

Using the perfect gas law, $p = (R/\mu)\rho T$, we can derive the density and temperature ratios across the shock

$$\frac{\rho_2}{\rho_1} = \frac{u_1}{u_2} = \frac{u_1^2}{u_1 u_2} = \frac{u_1^2}{c_s^{*2}} = \frac{(\gamma + 1)M_1^2}{(\gamma - 1)M_1^2 + 2}, \quad (M_1 = u_1/c_{s1}),$$

$p_2 - p_1 = \rho_1 u_1(u_1 - u_2)$, since $\rho_1 u_1 = \rho_2 u_2$, which gives

$$\frac{p_2}{p_1} - 1 = \frac{\rho_1 u_1^2}{p} \left(1 - \frac{u_2}{u_1} \right) = \frac{2\gamma(M_1^2 - 1)}{\gamma + 1}. \tag{12.21}$$

Using the perfect gas law, $p = (R/\mu)\rho T$, we summarize the jumps across the shock

$$\frac{\rho_2}{\rho_1} = \frac{(\gamma + 1)M_1^2}{(\gamma - 1)M_1^2 + 2} = \frac{u_1}{u_2},$$

$$\frac{p_2}{p_1} = \left(\frac{2\gamma}{\gamma + 1} \right)^{M_1^2 - 1} \quad \text{and}$$

$$\frac{T_2}{T_1} = \frac{p_2}{p_1} \frac{\rho_1}{\rho_2} = 2\gamma(M_1^2 - 1) \left\{ \frac{(\gamma - 1)M_1^2 + 2}{(\gamma + 1)^2 M_1^2} \right\}. \tag{12.22}$$

Again, from the Rankine-Hugoniot relations, we obtain

$$\rho_1 u_1 = \rho_2 u_2, \quad p_1 + \rho_1 u_1^2 = p_2 + \rho_2 u_2^2, \text{ and}$$

$$\frac{u_1^2}{2} + \frac{\gamma}{\gamma - 1}\frac{p_1}{\rho_1} = \frac{u_2^2}{2} + \frac{\gamma}{\gamma - 1}\frac{p_2}{\rho_2} \text{ which gives}$$

$$u_1 + \frac{p_1}{\rho_1 u_1} = u_1 + \frac{p_1}{\rho_1 u_1},$$

$$u_2 - u_1 = \frac{p_1 - p_2}{\rho_1 u_1}. \tag{12.23}$$

Multiplying this equation by $(u_2 + u_1)$, we recover

$$u_2^2 - u_1^2 = \frac{p_1 - p_2}{\rho_1}\left(1 + \frac{u_2}{u_1}\right) = (p_1 - p_2)\left(\frac{1}{\rho_1} + \frac{1}{\rho_2}\right). \tag{12.24}$$

Using the energy equation, we have

$$\frac{2\gamma}{\gamma - 1}\left(\frac{p_1}{\rho_1} - \frac{p_2}{\rho_2}\right) = (p_1 - p_2)\left(\frac{1}{\rho_1} + \frac{1}{\rho_2}\right). \tag{12.25}$$

Solving these two equations we obtain the density ratio,

$$\frac{\rho_2}{\rho_1} = \frac{(\gamma - 1)p_1 + (\gamma + 1)p_2}{(\gamma + 1)p_1 + (\gamma - 1)p_2} \tag{12.26}$$

and from the momentum conservation equation, we get

$$\frac{p_2}{p_1} - 1 = \frac{\rho_1 u_1}{p_1}\left(1 - \frac{u_2}{u_1}\right) = \frac{2\gamma M_1^2(M_1^2 - 1)}{(\gamma + 1)M_1^2}. \tag{12.27}$$

It is customary to define strength of the shock by the ratio, $(p_2 - p_1)/p_1$ which can become inordinately large across the shock, while the density jump tends to a finite ratio $(\gamma + 1)/(\gamma - 1)$ and the velocity jump, $u_2/u_1 \to (\gamma - 1)/(\gamma + 1)$. The temperature jump, t_2/T_1, however, can become indefinitely large.

Invoking the energy equation in the form, $I_1 + u_1^2/2 = I_2 + u_2^2/2$, we can deduce the enthalpy across the shock, $I_1 - I_2 = u_2^2/2 - u_1^2/2$ with $\rho_1 u_1 = \rho_2 u_2$. Using the momentum equation, $p_1 + \rho_1 u_1^2 = p_2 + \rho_2 u_2^2$, we obtain

$$\frac{1}{2}u_2^2 - \frac{\rho_1}{2\rho_2}u_1^2 = \frac{p_1}{2\rho_2} - \frac{p_2}{2\rho_2}, \text{ or}$$

$$\frac{1}{2}u_1^2 - \frac{\rho_2}{2\rho_1}u_2^2 = \frac{p_2}{2\rho_1} - \frac{p_1}{2\rho_1}, \tag{12.28}$$

thus giving

$$\frac{1}{2}u_2^2 - \frac{1}{2}u_1^2 - \frac{1}{2}\left(\frac{\rho_2}{2\rho_1}u_2^2 - \frac{\rho_1}{2\rho_2}u_1^2\right) = \frac{1}{2}\left(\frac{p_1}{\rho_2} - \frac{p_2}{\rho_2} + \frac{p_1}{\rho_1} - \frac{p_2}{\rho_2}\right). \qquad (12.29)$$

This implies

$$\Delta I = \frac{1}{2}\left(\frac{1}{\rho_1} + \frac{1}{\rho_2}\right)\Delta p = 0 \qquad (12.30)$$

since $\rho_1 u_1 = \rho_2 u_2$.

This expression for enthalpy jump can be extended to magnetohydrodynamic shocks with the magnetic fields parallel to the shock wave, a situation when two masses of magnetized plasma collide with a relative velocity large compared to sound wave and Alfven wave, exciting the shock wave. Assume the shock is reduced to rest in a chosen coordinate frame. The flow-field outside is supposed to be uniform outside the shock and so is the magnetic field. There are no currents outside the shock which is itself a current sheet. The only gradients are in the x-direction, say. Then, MHD equations give curl $\mathbf{E} = 0$ implying that $\partial E_z/\partial x = 0$, or, E_z is a constant, i.e., $E_{z1} = E_{z2}$ across the shock. Likewise, Ohm's law gives $\mathbf{E}+\mathbf{u}\times\mathbf{B}/c = 0$, or $[E_z+u_{1x}B_{1y}/c]$, jump across the shock , $= 0$, or $u_1 B_{1z} = u_2 B_{2z}$. Using the mass conservation equation, $\rho_1 u_1 = \rho_2 u_2$, we obtain $B_{1z}/\rho_1 = B_{2z}/\rho_2$, or, the entity (B/ρ) is invariant across the shock.

Appealing to the equations of momentum and energy conservation in the presence of magnetic field, we can write

$$p_1 + \rho_1 u_1^2 + \frac{B_1^2}{8\pi} = p_2 + \rho_2 u_2^2 + \frac{B_2^2}{8\pi}, \text{ and}$$

$$\frac{u_1^2}{2} + \frac{p_1}{\rho_1} + E_1 + \frac{B_1^2}{4\pi\rho_1} = \frac{u_2^2}{2} + \frac{p_2}{\rho_2} + E_2 + \frac{B_2^2}{4\pi\rho_2}. \qquad (12.31)$$

In terms of specific enthalpy, $I = E + p/\rho$,

$$I_1 - I_2 + \frac{1}{2}(u_1^2 - u_2^2) + \left(\frac{B_1^2}{4\pi\rho_1} - \frac{B_2^2}{4\pi\rho_2}\right) = 0, \text{ or}$$

$$\Delta I - \frac{V_1 + V_2}{2}\Delta p - \frac{(V_1 + V_2)}{16\pi}(B_1^2 - B_2^2) + \frac{B_1^2 V_1 - B_2^2 V_2}{4\pi} = 0, \qquad (12.32)$$

which may be written as

$$\Delta I - \frac{V_1 + V_2}{2}\Delta p + \frac{(\Delta B)^2 \Delta V}{16\pi} = 0, \qquad (12.33)$$

Fig. 12.1 After a spherically symmetric explosion, a blast wavefront is propagating with a velocity u at a time t, the radius is $R(t)$

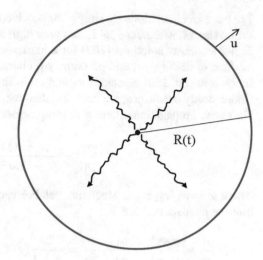

known as Lust's equation. Here, $V = 1/\rho$ is the specific volume. Rewriting $p^* = p + B^2/8\pi$ and $I^* = I + B^2/4\pi\rho$, for a fictitious gas designated by properties p^*, I^*, ρ, T, s, then the relation $T dS = dI^* - dp^*/\rho$ holds, and the above equation becomes $\Delta I^* - (V_1 + V_2)\Delta p^*/2 = 0$, which is equivalent to non-magnetic case, $\Delta I - (V_1 + V_2)\Delta p/2 = 0$.

Exercise Estimate the thickness of shock waves [54] (Fig. 12.1).

12.2 Blast Waves

When a powerful explosion occurs in a gaseous medium, it gives rise to an intense disturbance in the vicinity of the explosion such as an atomic device, supernova explosion which ploughs through the surrounding medium, decreasing in intensity with time. The front of such a propagating disturbance is called a blast wave with associated radius, $R(t)$. For spherically symmetric explosion, for $r > R, u = 0, p = p_0, \rho = \rho_0, T = T_0, c_s = c_{s0}$. Neglecting viscous effects and thermal conduction and the effects of gravity, the equations of motion in the region, $r < R$, can be written as

$$\text{mass conservation: } \frac{\partial \rho}{\partial t} + u \frac{\partial \rho}{\partial r} + \rho \left(\frac{\partial u}{\partial r} + \frac{2u}{r} \right) = 0,$$

$$\text{momentum conservation: } \frac{\partial u}{\partial t} + u \frac{\partial u}{\partial r} = -\frac{1}{\rho} \frac{\partial p}{\partial r},$$

$$\text{energy conservation: } \frac{ds}{dt} = 0, \text{ i.e. } \frac{\partial}{\partial t} \left(\frac{p}{\rho^\gamma} \right) + u \frac{\partial}{\partial r} \left(\frac{p}{\rho^\gamma} \right) = 0 \qquad (12.34)$$

The boundary conditions are simply the shock condition to be satisfied at $r = R$. For strong shocks, with $u/c_{s0} \gg 1$, i.e., very high Mach number, we apply the Sedov-Taylor similarity solutions [102] for a blast wave propagating into an undisturbed medium of density, ρ_1, and pressure, p_1, choosing a reference frame in which the shock is at rest, with velocities u_1 and u_2 in the two regions 1 and 2 respectively. Under steady conditions, we have the Rankine-Hugoniot equations controlling the blast wave propagation. These give jump across the shock,

$$\frac{\rho_2}{\rho_1} = \frac{(\gamma + 1)M^2}{(\gamma - 1)M^2 + 2}, \tag{12.35}$$

where $M = u_1/c_{s1}$ is the Mach number. For very strong shock (or blast) waves, the limiting jumps across are

$$\frac{\rho}{\rho_0} = \frac{\gamma + 1}{\gamma - 1} = \frac{u_1}{u_2}, \quad \frac{p}{p_0} = \frac{2\gamma}{\gamma + 1}(u/c_{s0})^2, \quad \frac{u}{U} = \frac{2}{\gamma + 1} \tag{12.36}$$

at the shock wave boundary (Fig. 12.2).

For self-similar solutions in which all the variables are function of $\eta = r/R$, the boundary corresponding to $\eta = 1$, we choose $\rho/\rho_0 = \psi(\eta)$ where $\psi(1) = (\gamma + 1)/(\gamma 1)$; $p/p_0 = (U^2/c^2)f(\eta)$, where $f(1) = 2\gamma/(\gamma + 1)$; $u/U = \phi(\eta)$, where $\phi(1) = 2/(\gamma + 1)$.

The energy balance consideration demands the energy inside the shock to equal the energy released by explosion

$$E = \int_0^R 4\pi r^2 dr \rho \left(\frac{u^2}{2} + c_V T \right), \tag{12.37}$$

Fig. 12.2 For strong shocks at high Mach number, similarity solutions apply for a propagating blast wave. A reference frame is chosen where shock is at rest in the two regions, 1 and 2

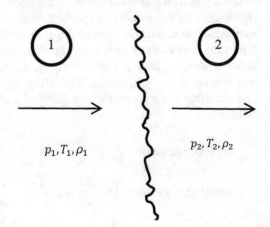

where $u^2/2$ is the kinetic energy and $c_v T$ is the internal energy per unit mass. Hence,

$$\frac{E}{\rho_0} = \int_0^1 4\pi R^3 \eta^2 U^2 d\eta \left(\psi\, \phi^2 + \frac{1}{\gamma(\gamma-1)} \right) = R^3 U^2 A, \text{ where}$$

$$A = \int_0^1 4\pi \eta^2 d\eta \left(\frac{1}{2}\psi\, \phi^2 + \frac{f}{\gamma(\gamma-1)} \right), \tag{12.38}$$

since ψ, ϕ, and f are all functions of η alone, when integrated over $\eta \in (0, 1)$. A turns out to be a constant. Thus, $U^2 R^3 = E/A\rho_0$ is a constant, independent of time, t. If we now take $U \propto R^n$, with $U = dR(t)/dt$, then $U^2 R^3 \sim R^{2n+3} = \text{constant} = (1/(n-1))$ implies $n = -3/2$ and use of $U = dR(t)/dt$ gives $R \sim t^{2/5}$ and $U \sim t^{-3/5}$. This behaviour of the self-similar evolution also follows from scaling arguments, i.e., the blast wave evolves in such a manner if some initial configuration expands uniformly with the subsequent configuration appearing like an enlargement of the earlier configuration. Hence, on dimensional grounds, $R = (Et^2/\rho_1)^{1/5}$ is a scale parameter designating size of the blast wave at time t after explosion. Introducing a dimensionless parameter $\eta = r/R$ or

$$\eta = r \left(\frac{\rho_1}{Et^2} \right)^{1/5} \sim t^{-2/5}. \tag{12.39}$$

In summary, the Rankine-Hugoniot conditions give

$$\rho_1 = \rho_0 \left(\frac{\gamma+1}{\gamma-1} \right), \quad p_1 = \rho_0 \left(\frac{2}{\gamma+1} \right) U^2, \quad u_1 = \frac{2U}{\gamma+1}, \tag{12.40}$$

where $U = dR(t)/dt$ is velocity of shock front. Suppose energy $E\ (\sim Mu^2)$ is suddenly released in an explosion producing a blast wave which progresses into an ambient medium of density ρ_0. Assume the pressure p_0 in the ambient medium is negligible compared to the pressure p inside the blast wave, i.e. $p_0 \simeq 0$.

Let R be the scale parameter characterizing size of blast wave at time t after the explosion. Apart from being a monotonically increasing function of time t, the evolution of $R(t)$ must also depend on E and ρ_0. Now the only way of combining $E\ (\simeq Mu^2)$, ρ_0 and t to obtain a dimension of length is $R \sim (Et^2/\rho_0)^{1/5}$ where $E \sim \rho U^2 R^3$. Let $r(t)$ be the radius of a shell of gas inside the spherical blast wave. For a self-similar explosion, $r(t)$ has to evolve in the same way as R, so that we can introduce a dimensionless parameter, $\eta = r/R = r(\rho_0/Et^2)^{1/5}$, such that the value of η for a particular shell of gas does not change with time t, and each shell may be specified by a particular value of η. Let η_0 correspond to the shock front so that the radius of the spherical blast is given by $r_s(t) = \eta_0 (Et^2/\rho_0)^{1/5}$ and the expansion velocity,

$$v_s(t) = \frac{dr_s(t)}{dt} = \frac{2}{5} \frac{r_s}{t} = \frac{2}{5}\eta_0 \left(\frac{E}{\rho_0 t^3} \right)^{1/5}. \tag{12.41}$$

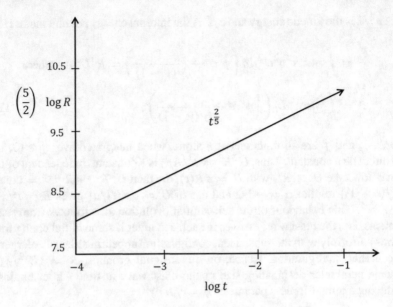

Fig. 12.3 The size of the spherical blast wave increases with time as $t^{2/5}$

In short, the size of the spherical blast wave thus increases as $R \sim t^{2/5}$ (Fig. 12.3) and velocity of the front declines as $U \sim t^{-3/5}$. This was discovered by Taylor [115] who worked out the propagation profile of the blast wave produced by the US atomic explosion.

Another example is a typical supernova explosion resulting from the evolution of a massive star ejecting over $10M_\odot$ with an initial velocity of $10,000$ km/s, so that the energy input by the supernova is, $E \sim Mu^2 \sim 10^{51-52}$ erg. The blast wave expands into the surrounding interstellar medium having typical density of $n \sim 1\,\mathrm{cm}^{-1}$, or, $\rho \sim 10^{-24}$ g/c.c. Taking ρ to be of order unity, we derive

$$r_s(t) \simeq 0.3t^{2/5}\,pc \tag{12.42}$$

and $u_s(t) \simeq 10^5 t^{-3/5}$ km/s, with t measured in years [54].

12.3 Structure of a Shock

Recall the stress-strain relationship,

$$\sigma_{ij} = -p\delta_{ij} + 2\mu_s\left(e_{ij} + \frac{1}{3}\Delta\delta_{ij}\right) + \eta_b\delta_{ij}\Delta, \tag{12.43}$$

where $\Delta = \partial u_i/\partial x_i = \text{div }\boldsymbol{u}$, μ_s is the coefficient of shear viscosity and η_b, that of bulk viscosity. Appealing to the first law of thermodynamics,

$$TdS = dE + pd(1/\rho), \qquad V = 1/\rho \tag{12.44}$$

and hence the energy equation

$$\rho T \frac{dS}{dt} = \rho \frac{dE}{dt} + p\rho \frac{d}{dt}\frac{1}{\rho}$$

$$= \sigma_{ij} e_{ij} + p\Delta + \frac{\partial}{\partial x_i}\left(\kappa \frac{\partial T}{\partial x_i}\right). \tag{12.45}$$

Using the mass conservation equation,

$$\frac{d\rho}{dt} + \rho \text{div }\boldsymbol{u} = 0, \text{ or } \frac{d\rho}{dt} + \rho\Delta = 0, \text{ we recover}$$

$$p\rho \frac{d}{dt}\frac{1}{\rho} = -\frac{p}{\rho}\frac{d\rho}{dt} = p\Delta \tag{12.46}$$

to write the energy equation in the form

$$\rho \frac{dE}{dt} = \sigma_{ij} \frac{\partial u_i}{\partial x_j} + \frac{\partial}{\partial x_i}\left(\kappa \frac{\partial T}{\partial x_i}\right). \tag{12.47}$$

We can then express the energy equation as,

$$\rho T \frac{dS}{dt} = \sigma_{ij} e_{ij} + p\Delta + \frac{\partial}{\partial x_i}\left(\kappa \frac{\partial T}{\partial x_i}\right)$$

$$= 2\mu_s\left(e_{ij} - \frac{1}{3}\Delta^2\right) + \eta_b\Delta^2 + \frac{\kappa}{T}\left(\frac{\partial T}{\partial x_i}\right)^2$$

$$= \Phi_1 + \Phi_2 + T\frac{\partial}{\partial x_i}\left(\kappa \frac{\partial T}{\partial x_i}\right). \tag{12.48}$$

$(\Phi_1 + \Phi_2)$ evidently represents the entropy generation on account of energy dissipation by viscosity an thermal conductivity.

On dividing by T, and integrating over volume τ, we obtain

$$\frac{d}{dt}\int \rho S d\tau = \int_\tau \frac{\Phi_1 + \Phi_2}{T} d\tau + \int_S \left(\kappa \frac{\partial T}{\partial x_i}\right)dS_i. \tag{12.49}$$

12.3.1 Structure of Weak Shocks

Let us consider the structure of weak shocks in the case of one-dimensional steady flows. Equations of motion take the following form:

1. Momentum conservation:

$$\rho u \frac{du}{dx} = -\frac{dp}{dx} + \frac{d}{dx}\left(\bar{\mu}\frac{du}{dx}\right). \tag{12.50}$$

2. Continuity equation:

$$\frac{d}{dx}(\rho u) = 0. \tag{12.51}$$

3. Energy equation:

$$\rho u \frac{d}{dx}\left(I + \frac{1}{2}u^2\right) = \frac{d}{dx}\left(\kappa\frac{dT}{dx}\right) + \frac{d}{dx}\left(\bar{\mu}u\frac{du}{dx}\right). \tag{12.52}$$

Here, $I = E + p/\rho$ denotes the enthalpy, κ is the thermal conductivity and $\bar{\mu}$ the viscosity. Th
4. Boundary conditions:

$$u \to u_1, \text{ as } x \to -\infty, \quad u \to u_2, \text{ as } x \to +\infty. \tag{12.53}$$

These equations can be solved numerically if the coefficients κ and $\bar{\mu}$ are prescribed as functions of state variables. The first integrals of these equations are:

$$\rho u = m,$$

$$mu = -p + \bar{\mu}\frac{du}{dx} + mu_1 + p_1,$$

$$m\left(I + \frac{1}{2}u^2\right) = m\left(I_1 + \frac{1}{2}u_1^2\right) + \kappa\frac{dT}{dx} + \bar{\mu}u\frac{du}{dx}. \tag{12.54}$$

12.3.2 Estimation of Shock Thickness

Rate of entropy production,

$$m\Delta S = m(S_2 - S_1)$$

$$= \int \left\{\frac{\bar{\mu}}{T}\left(\frac{du}{dx}\right)^2 + \frac{\kappa}{T^2}\left(\frac{dT}{dx}\right)^2\right\}dx$$

$$> \int_0^h \left\{ \frac{\bar{\mu}}{T} \left(\frac{du}{dx} \right)^2 + \frac{\kappa}{T^2} \left(\frac{dT}{dx} \right)^2 \right\} dx$$

$$\geq \frac{1}{h} \left[\int_0^h \left(\frac{\bar{\mu}}{T} \right)^{1/2} \frac{du}{dx} dx \right]^2 + \frac{1}{h} \int_0^h \int_0^h \frac{\kappa^{1/2}}{T} \left(\frac{dT}{dx} \right) dx$$

$$\geq \frac{1}{h} \left(\frac{\bar{\mu}}{T} \right) (u_1 - u_2)^2 + \frac{1}{h} \left\{ \int_{T_1}^{T_2} \frac{\kappa^{1/2}}{T} dT \right\}^2 . \tag{12.55}$$

Hence, the shock thickness,

$$h > \left(\frac{\bar{\mu}_2}{T_2} \right)^2 (u_1 - u_2)^2 + \left\{ \int_{T_1}^{T_2} \frac{\kappa^{1/2}}{T} dT \right\}^2 \equiv \rho_1 u_1 \Delta S. \tag{12.56}$$

For a Mach number, $M \simeq 2(10)$, $h \simeq 2 \times 10^{-4}(10^{-4})$ cm. In the case of weak shocks,

$$\Delta S = c_V \frac{\gamma(\gamma^2 - 1)}{12} \left(\frac{\Delta U}{U} \right)^3, \tag{12.57}$$

by using thermodynamic relations for perfect gas,

$$s_2 - s_1 = c_V \log \frac{p_2}{p_1} - \gamma c_V \log \frac{\rho_2}{\rho_1}, \tag{12.58}$$

which implies

$$s_2 - s_1 = \frac{2\gamma(\gamma - 1)}{3(\gamma + 1)^2} (M_1^2 - 1)^3 = \frac{\gamma^2 - 1}{12\gamma^2} \left(\frac{p_2 - p_1}{p_1} \right)^3$$

$$= \frac{12\delta}{(\gamma + 1)\Delta u} \tag{12.59}$$

where $\delta = 4v_s/3 + v_b + \kappa(\gamma - 1)$.

12.3.3 Velocity Profile Through a Weak Shock

For a weak shock, the viscosity and thermal conductivity do not vary much across the shock. We also assume c_P, c_V to be constants. For a perfect gas, pressure $p =$

$(c_P - c_V)\rho T$ and enthalpy, $I = c_P T, \overline{\mu} = 4\mu_s/3 + \eta_b$. Conservation equations can be cast in the following form:

1. Mass: $\rho u = m$.
2. Momentum:

$$mu = -p + \overline{\mu}\frac{du}{dx} + \text{constant}, \tag{12.60}$$

which goes into

$$mu(u - u_1) + m(c_P - c_V)T - m(c_P - c_V)\frac{uT_1}{u_1} = \overline{\mu}u\frac{du}{dx}. \tag{12.61}$$

In this expression, RHS $= \overline{\mu}u\Delta u/h$ and the first term on the LHS is $\rho u^2 \Delta u$. Hence,

$$\frac{\text{RHS}}{\text{LHS}} = \frac{\overline{\mu}}{\rho u h} = \frac{v}{uh} \ll 1, \tag{12.62}$$

since $h\Delta u/v = O(1)$ and $\Delta u/u \ll 1$ for weak shocks. Note

$$\frac{v}{uh} = \frac{v}{(\Delta u)h}\frac{\Delta u}{u} \simeq \frac{\Delta u}{u} \ll 1. \tag{12.63}$$

To the first order approximation, $c_P T + u^2/2$ is a constant from energy equation. Hence, $c_P dT/dx = -udU/dx$, the energy and momentum equations then become

$$mc_P(T - T_1) + \frac{m}{2}(u^2 - u_1^2) = -\frac{\kappa_1 u_1}{c_P}\frac{du}{dx} + \overline{\mu}u_1\frac{du}{dx},$$

$$mu(u - u_1) + m(c_P - c_V)\left(T - \frac{uT_1}{u_1}\right) = \overline{\mu}u_1\frac{du}{dx}. \tag{12.64}$$

Multiplying the first of these equations by $(\gamma - 1)$ and the second by γ and subtracting, we recover

$$\overline{\mu}u_1\frac{du}{dx} + \frac{\gamma - 1}{c_P}\kappa_1 u_1\frac{du}{dx} = \rho_1\delta_1 u_1\frac{du}{dx}, \text{ where } \delta = 4v/3 + v_b + \kappa(\gamma - 1)$$

$$= \frac{mu^2}{2}(\gamma + 1) + Au + B$$

$$= -\frac{m}{2}(\gamma + 1)(u_1 - u_2)(u - u_2), \tag{12.65}$$

since LHS = 0 at $u = u_1, u = u_2$, and at ∞, all the gradients on LHS vanish. Hence,

$$\frac{du}{(u_1 - u)(u - u_2)} = -\frac{m(\gamma + 1)}{2\rho_1 \delta_1 u_1} dx = -\frac{(\gamma + 1)}{2\delta_1} dx \tag{12.66}$$

which integrates to give

$$\log \frac{u - u_2}{u_1 - u} = -\frac{(\gamma + 1)}{2\delta_1} \frac{x}{u_1 - u_2} = \frac{x}{h'}, \text{ say} \tag{12.67}$$

which gives the velocity profile

$$\frac{u - u_2}{u_1 - u} = e^{-x/h'}, \text{ or}$$

$$u = \frac{u_1 + u_2}{2} - \frac{u_1 - u_2}{2} \tanh\left(\frac{x}{2h'}\right) \tag{12.68}$$

Taylor thickness, h_T is defined as the distance in which $(u - u_2)/(u_1 - u)$ goes from 0.05 to 20. This gives

$$\frac{\gamma + 1}{2\delta_1 (u_1 - u_2)} h_T = 2 \log 20, \text{ or}$$

$$h_T = \frac{11.98 \delta_1}{(\gamma + 1)(u_1 - u_2)} \tag{12.69}$$

where $\delta_1 = 4\nu_1/3 + \nu_{b1} + \kappa_1(\gamma - 1)$.

Exercise 12.1 For a given shock strength and angle of incidence what reflected waves will leave the gas flowing parallel to the wall?

Astrophysical Fluid Mechanics

13

13.1 Accretion onto Compact Objects

Cosmic bodies are endowed with intrinsic energies such as gravitational, thermal, nuclear, rotational and magnetic field components. There is, however, also an extrinsic source arising from the release of gravitational energy released by the in-falling material onto the deep gravitational well, provided by compact objects. The role of collapsed objects in attracting neighbouring material onto themselves and effectively generating radiation from the gravitational potential energy released was highlighted by Zeldovich's group in 1979s, after the discovery of cosmic X-ray and γ-ray sources. Thus, should the material arriving at stellar surfaces, e.g. white dwarfs, neutron stars and black holes were to be thermalized, the flow field would be characterized by motions close to the free-fall velocity of order,

$$\frac{1}{2}v_{\mathrm{ff}}^2 = \frac{GM}{R}, \ \text{ or } \ v_{\mathrm{ff}} = 1.6 \times 10^{10} (\mathrm{cm\,s}^{-1}) \left(\frac{M}{M_\odot}\right)^{1/2} \left(\frac{R}{10^6 \ \mathrm{cm}}\right)^{1/2} \tag{13.1}$$

and the temperature of order, $T_{\mathrm{ff}} = \frac{GM}{R}\frac{m_H}{k_B}$, will be reached. Here, M is the typical mass of the compact object and R, its radius.

The accompanying comparative table summarizes the kinetic energy of the in-fall, efficiency of gravitational energy, $\epsilon = GM/Rc^2$, and the resulting luminosity, $L = (GM/R)\dot{M}$ where $\dot{M}$ is the mass capture rate (i.e., $L = \epsilon \dot{M}c^2 \ \mathrm{erg/s}$). In Table 13.1, $\dot{M} = 10^{-8} \ M_\odot/\mathrm{year}$.

The process of infall of material onto a collapsed object evidently plays a vital role in generating high energy events and is clearly important in the context of high-energy astrophysics. This is largely because when matter descends onto a Sun-like star, the gravitational energy liberated is only about a millionth of the rest mass energy. But if the same amount of material falls onto the much deeper potential well associated with a collapsed object such as the neutron star or the black hole as much as 10–50% of the rest mass energy could be converted into radiation.

© The Author(s), under exclusive license to Springer Nature Switzerland AG 2022
S. R. Jain et al., *A Primer on Fluid Mechanics with Applications*,
https://doi.org/10.1007/978-3-031-20487-6_13

Table 13.1 For some stars and black hole, we compare the maximum kinetic energy in a free fall, and the luminosity generated

Object	Max. KE of infall at the surface	Efficiency (GM/Rc^2)	Luminosity generated $GM\dot{M}/R$ (erg/s)
Sun-like star	1 KeV	10^{-6}	10^{32}
White dwarf	100 KeV	10^{-4}	10^{35}
Neutron star	≤ 100 MeV	10^{-2}	10^{38}
Black hole	~ 100 MeV	0.06–0.4	10^{38}

Let us illustrate the classical solution (also referred to as Bondi's solution[17]) for spherical symmetric accretion onto a compact body of mass, M and radius R at rest in the medium whose thermal yield at a large distance from the body is denoted by p_∞, ρ_∞, T_∞ and $c_{s\infty}$, assuming the flow to be adiabatic, the pressure-density relationship may be written as

$$\frac{p}{p_\infty} = \left(\frac{\rho}{\rho_\infty}, \right)^\gamma \quad \text{and} \quad c_s = \sqrt{\frac{\gamma p}{\rho}}. \tag{13.2}$$

Measuring the radial coordinate r from the centre of the accreting body and denoting by u, the inward flow velocity of the gas, the governing equations for the spherically symmetric flow field can be written as

1. mass conservation:

$$4\pi r^2 \dot{\rho} = \text{constant} = \dot{M}, \quad \text{mass accretion rate,}$$

2. momentum conservation:

$$\rho u \frac{du}{dr} = \frac{dp}{dr} - \frac{GM}{r^2}\rho, \tag{13.3}$$

3. Energy conservation (Bernoulli equation):

$$\frac{1}{2}u^2 + \frac{c_s^2}{\gamma - 1} - \frac{GM}{r} = \frac{c_{s\infty}^2}{\gamma - 1}, \tag{13.4}$$

noting at $r = \infty$, $u = 0$ and $c_s = c_{s\infty}$, with $c_{s\infty} = (\gamma p_\infty / \rho_\infty)^{1/2}$.

Differentiating the mass conservation equation w.r.t. r and inserting the expression for du/dr from the momentum equation, we recover,

$$\frac{c_s^2 - u^2}{c_s^2} \frac{1}{\rho} \frac{dp}{dr} = \frac{2u}{r} - \frac{GM}{r^2} \tag{13.5}$$

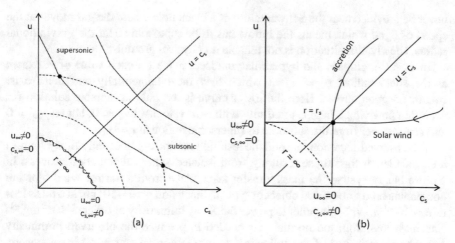

Fig. 13.1 Bondi's solution are shown here schematically in the $u - c_s$-plane

which determines the radius at which the sonic velocity, c_s is reached, viz., $u = c_s$ occurs at $r = r_s = GM/2c_s^2$, the sonic radius.

Let us consider the two equations:

$$u = \frac{\dot{M}}{4\pi r^2 \rho_\infty} \left(\frac{c_{s\infty}}{c_s}\right)^{2/(\gamma-1)} \quad \text{, and}$$

$$\frac{1}{2}u^2 + \frac{c_s^2}{\gamma - 1} - \frac{GM}{r} = \text{constant} = \frac{c_{s\infty}^2}{\gamma - 1} \quad (13.6)$$

Bondi discussed the elegant solution of these equations in the $(u - c_s)$-plane (Fig. 13.1), where the Bernoulli equation for fixed r defines an ellipse. The different values of r correspond to different ellipses, and the mass conservation equation $uc_s^2/(\gamma - 1) = \text{constant}$ defines a hyperbola for each r. Thus, for every value of r, we have two equations for two variables, u and c_s. The solution of these equations would involve finding the intersection point of the ellipse and the hyperbola determining u and c_s for the particular radius, r. It is clear that for every pair of ellipse and hyperbola, there are either two intersection points, or, one point of tangency or no intersection at all. Evidently, for a particular chosen mass accretion rate, $\dot{M}$, there may not exist any value of r at which the curves intersect and the two equations are incompatible for that choice of the mass accretion rate, $\dot{M}$ and such flow fields cannot occur.

(a) Ellipses and hyperbolae intersect twice for every value of r and there are two separate curves, $u = u(c_s)$ corresponding to two families of intersection points, with the curves beginning on the innermost ellipse, $r = \infty$. In the (u, c_s)-plane, for case (a), the lower curve remains in the subsonic region throughout, which can happen only for a gas falling into a gravitating body that requires back pressure at all radii to resist the inflow to keep it subsonic. Clearly, such a back pressure can never be provided near the Schwarzchild of a black hole where the gas moves at the speed of light, which means the inflow has to be supersonic near the gravitational radius. Clearly, this situation is not tenable on physical grounds.

In case (b), ellipses and hyperbolae intersect twice for each value of r, except at the sonic radius, $r = r_s$ at which they meet tangentially and there exist two intersection curves. Here the lower curve is the only acceptable solution for accretion flow which begins at infinity with zero velocity, $u_\infty = 0$, but $(c_s)_\infty \neq 0$ and crosses over from the subsonic to supersonic region at $r = r_s$.

The observed accretion rate and the resulting luminosity of cosmic X-ray sources are much larger than those resulting from isolated spherical accretion process. In a close binary system the mass transfer takes place from a normal wind blowing companion onto a collapsed object of typical dimension $\sim 10^6$–10^7 cm at a mass loss rate $\sim 10^{-9} M_\odot yr^{-1}$, in order to power the X-ray luminosity of order $\sim 10^{38}$ erg^{-1}. The mass-loss from the normal star needed to produce the observed luminosity can take place in one of the following ways: (1) normal star in a binary system, during the course of its evolution fills the so-called Roche lobe and transfers the overflowing matter through the inner Lagrangian point; (2) sufficiently intense stellar wind from the massive companion can flow onto the compact object; (3) outer layer of a normal star gets heated by the X-ray source resulting in a self-excited wind.

Another feature of accreting gas in a binary system is endowed with a considerable amount of angular momentum and, as a result, instead of falling radially onto the compact objects, the gas orbits around the central gravitating source forming a thin accretion disk. An essential feature of accretion disk is the viscosity of the inflowing circulating gas which effectively plays a crucial role in effectively transporting angular momentum outwards. This results in braking of the inner regions of the accretion disk from where the gas, on losing its angular momentum swirls inwards closer to the compact object. There is, of course, the limit to the luminosity generated in such an accretion process. The resulting luminosity for a spherical accretion causes a radiative repulsion of an electron at the distance r from a star of magnitude, $(L/4\pi r^2 c)\sigma_T$ (σ_T being Thomson scattering cross-section) which has to counter the gravitational attraction on an infalling proton of magnitude, $(GM/r^2)m_H$, thus resulting in the so-called Eddington luminosity when the two forces balance each other, $L\sigma_T/4\pi r^2 = GMm_H/r^2$ to give

$$L_{Eddington} = \frac{4\pi GMcm_H}{\sigma_T} \simeq 10^4 \frac{L}{L_\odot}. \tag{13.7}$$

A partially ionized gas possesses a mechanism for dissipating the energy of electric currents which may be absent in a fully ionized medium. The force that drives the velocity of the electrons related to the ions also drives the velocity of the neutrals by the process of ambipolar diffusion of electrons and ions jointly relative to neutral atoms by the collision of neutrals with electrons and ions. The drag opposing the ion velocity, v_i exerted during collisions with neutral atoms, requires an increased energy supply to maintain the electric currents.

13.2   Magnetic Fields as Agents of Transporting Angular Momentum

Let us discuss the role of magnetic fields acting as agents for transporting angular momentum. Consider a star-forming gas cloud in a galaxy, collapsing as a result of Jeans instability bringing the angular momentum inherited from galactic rotation and threaded by the general magnetic field of the parent galaxy. Assuming the conservation of specific angular momentum per unit mass, we expect the angular momentum, Ωr^2 to be conserved and the angular velocity, Ω to keep rising with diminishing r. This will naturally lead to the situation of the angular velocity of the collapsing cloud being larger than that of the surrounding plasma. The magnetic field threading the collapsing cloud with the surrounding ambient plasma will resist such a variation of angular velocity and will tend to provide a magnetic braking mechanism to slow down the rapidly spinning collapsing cloud. Assume the surrounding plasma is co-rotating with the angular velocity, $\Omega(r)$ of the collapsing cloud upto a distance, r_Ω, and beyond this radius, the angular velocity is much smaller. The Maxwell stresses would be expected to spin up the plasma beyond the radial distance, $r = r_\Omega$ to approach the angular velocity Ω over time-scales needed to propagate the magnetic disturbances, with the Alfven speed $v_A = B/\sqrt{4\pi\rho}$. Thus, in a time interval, δt, a plasma shell between $r = r_\Omega$ and $r = r_\Omega + v_A\delta t$ is spun up to angular velocity, Ω. Noting the angular momentum associated per unit mass of the shell is $\frac{2}{3}\Omega r_\Omega^2$, so that the total angular momentum added to this shell $\frac{8\pi}{3}r_\Omega^4\rho v_A\delta t$, which can only be derived from the angular momentum supplied by the central star of mass, M, which is $\frac{2}{5}M\Omega r_\Omega^2$. Hence the angular momentum balance gives

$$\frac{2}{5}Mr_\Omega^2\frac{d\Omega}{dt} = -\frac{8\pi}{3}\Omega r_\Omega^4\rho\frac{B}{\sqrt{4\pi\rho}},\tag{13.8}$$

thus we have an approximate equation for the magnetic braking of the central object. The collapse in the presence of angular momentum is seldom spherically symmetric and this leads to the formation of disks with the centrifugal force balancing the gravitational attraction near the equatorial latitudes. Let us then consider a simplest possible magnetic configuration with the axis of rotation in the w-direction and is independent of z and θ, viz., $\boldsymbol{B} = (B_r, B_\theta, B_z)$ in the cylindrical geometry, i.e., $\boldsymbol{B} = (B_0/w, B_\theta, B_z)$ satisfying the equation div $\boldsymbol{B} = 0$ for $\boldsymbol{u} = (u_w, u_\theta, u_z)$. The θ-component of the induction equation for a perfectly conducting fluid, $\partial\boldsymbol{B}/\partial t = $ curl $(\boldsymbol{u} \times \boldsymbol{B})$ then gives

$$\frac{\partial B_\theta}{\partial t} = B_0\frac{\partial\Omega}{\partial r} - \frac{\partial}{\partial r}(u_w B_\theta),\tag{13.9}$$

where surrounding the cylinder there is an envelope of low mass rotating with angular velocity Ω. The θ-component of the equation of motion gives

$$\rho u_w \frac{\partial}{\partial w}(\Omega w^2) = w\left(\frac{\text{curl } \boldsymbol{B} \times \boldsymbol{B}}{4\pi}\right) = w\frac{B_0}{4\pi w}\frac{\partial}{\partial w}(wB_\theta) \tag{13.10}$$

where ρ is the density of the envelope. Integrating these equations between two radii, w_1, w_2, we obtain

$$\frac{B_0}{2}[w \, B_\theta]_{w_1}^{w_2} = [2\pi\rho w \, u_w \, \Omega w^2]_{w_1}^{w_2} + \frac{\partial}{\partial t}\int_{w_1}^{w_2} \omega \, w^2 \, (2\pi\rho w \,)dw. \tag{13.11}$$

The LHS stands for the net transport of angular momentum into the volume by Maxwell's stresses, noting that the shearing stress transmitted along the magnetic line of force, B_w, as a consequence of twisting of the magnetic field is $B_w \, B_\theta/4\pi = B_0 \, B_\theta/4\pi w$. The first term on the RHS represents the net rate t which angular momentum is transported by moving matter out of the volume of unit height defined by radii, w_1, w_2, while the second term on the RHS stands for rate of increase of angular momentum in the volume through the density increases. This treatment of angular momentum transport is essentially developed by Mestel [76, 77].

We now illustrate the mechanism why the rotation of a cosmic cloud is reduced by the large-scale magnetic field. For this purpose, consider a cloud rotating relative to the surrounding interstellar medium with angular velocity, Ω and is coupled to it by a frozen-in magnetic field via changes induced in the azimuthal component, B_θ, of the field, through the induction equation. In cylindrical geometry, the induction equation, $\partial \boldsymbol{B}/\partial t = \text{curl }(\boldsymbol{u} \times \boldsymbol{B})$, for a perfectly conducting fluid, gives

$$\frac{\partial B_\theta}{\partial t} = w(\boldsymbol{B}.\nabla)\Omega, \tag{13.12}$$

and the θ-component of the equation of motion yields

$$\rho\frac{\partial}{\partial t}(\Omega w^2) = \boldsymbol{B}.\nabla\left(\frac{wB_\theta}{4\pi}\right). \tag{13.13}$$

Thus, the θ-component of the magnetic field, B_θ helps in re-distributing the angular momentum of the cloud. Combining the forgoing two equations produces a wave-like equation,

$$4\pi\rho\frac{\partial^2\Omega}{\partial t^2} = \frac{1}{w^2}\boldsymbol{B}.(w^2\boldsymbol{B}.\nabla\Omega). \tag{13.14}$$

In the simplest case of the magnetic field, B parallel to the angular momentum vector, $B \parallel \Omega$ for a uniform poloidal magnetic field directed along the z-axis, we recover the wave equation,

$$\frac{\partial^2 \Omega}{\partial t^2} = \frac{B^2}{4\pi\rho} \frac{\partial^2 \Omega}{\partial z^2}. \tag{13.15}$$

Thus, the rotation of the gas cloud relative to the background creates propagating magnetic disturbances resembling Alfvèn waves.

13.3 Origin and Maintenance of Cosmic Magnetic Fields

The origin of magnetic fields as well as rotation, i.e. angular momentum of the Earth and cosmic bodies is one of the challenging problems in astrophysics and cosmology. The pressure of permanent magnetism is clearly ruled out because of the high temperatures associated with the interior of cosmic bodies. There are two broad class of theories proposed to account for the source of cosmic magnetic fields:

1. *Fossil theory*: The observed magnetic field is supposed to be "fossil" relic of the original magnetic field inherited from the parent galaxy, within which the star was born as a result of the collapse of a gas cloud.
2. Suppose a protostar condenses from interstellar medium where the magnetic field of the order of micro-Gauss is pre-existing. The gravitational collapse as a result of Jeans' instability, increases the mean density by a factor of order $\sim 10^{18}$ as the radius decreases in ratio $10^6 : 1$. If the magnetic field is taken to be 'frozen' in the ionized plasma during the process of collapse, the magnetic field strength would rise from $\sim 10^{-6}$ G to 10^6 G which is much larger than the fields normally observed in ordinary stars. Clearly, this allows for comfortable margin for dissipation during the collapse as well as permitting the role of the neutrals present in the medium along with the ionized components [108]. Such a large magnetic field would be strong enough to exert a magnetic pressure, $\frac{B^2}{8\pi}$ to prevent the gravitational forces from leading to the formation of the star.

The "fossil" theory has other difficulties. These concern the decay time of stellar magnetic fields, assuming the stellar material to be at rest, i.e. $u = 0$ in the induction equation,

$$\frac{\partial B}{\partial t} = \operatorname{curl}(u \times B) + \frac{c^2}{4\pi\mu\sigma}\nabla^2 B \tag{13.16}$$

implies the estimated decay time of the order 10^{10} years. However, there are large-scale motions in a star which imply the ratio of diffusion time to convective time as $c^2/4\pi\mu\sigma U L$, using typical values for $\sigma \simeq 10^{16}\,\mathrm{s}^{-1}, U \simeq 10^2\,\mathrm{cm\,s}^{-1}, L \simeq 10^{10}\,\mathrm{cm}$ to get the ratio around $10^{-7} \ll 1$.

The "fossil theory" is not applicable to Earth, for example, because of the relatively short, $\sim 10^5$ years decay time. The theory has also other difficulties that it gives a sensibly constant field whereas fields of most magnetic stars are decidedly variable. There are, of course, other mechanisms possible to produce regular magnetic variations for the "fossil" field which we summarize in what follows.

The observations of Zeeman effect in the spectra of some stars indicate the presence of magnetic fields of sizeable strength ranging from several hundreds to tens of thousands gauss [5], which are often variable sometimes even reversing their direction. To account for underlying mechanism responsible for the magnetic variable stars there are number of proposals:

(a) *Magnetic oscillator theory*

In this a star is supposed to undergo periodic oscillations, driven by magnetic forces and other mechanical forces like rotation. The reversals in the observed magnetic polarity are supposed to arise from the interaction of torsional oscillations with the magnetic field.

(b) *Hydrodynamic cycle theory*

This mechanism is similar to the solar cycle theory which makes use of the observed differential rotation of the Sun whose equatorial regions faster than the polar latitudes. The initial poloidal magnetic field that connects the polar caps can be pulled out longitudinally by the differential rotation to produce a much stronger azimuthal field in the sub-surface regions. The conversion of the mechanical rotational energy into magnetic energy continues until a critical limit is reached when an instability sets in for the operation of magnetic buoyancy to lift the flux tubes to the surface resulting in the production of bipolar regions and sunspot groups [20]. As the cycle progresses it is hypothesized that the preceding bipolar regions drift towards the equator from both the hemispheres to get neutralized there by merging, while the following spots migrate towards the higher latitudes where the neutralization of the lines of force result.

Practically most of the mean field solar dynamo models hinge on the solution of the dynamo equation

$$\frac{\partial \boldsymbol{B}}{\partial t} + \text{curl}\,(\alpha \boldsymbol{B}) = v_m \nabla^2 \boldsymbol{B} - \text{curl}\,(\beta_{ij} \boldsymbol{B}) \qquad (13.17)$$

where the kinetic helicity, α and the diffusivity tensor, β_{ij} depend on the statistical properties of homogeneous turbulence, v_m denoting the magnetic diffusivity. This theory considers generation by dynamo action of magnetic fields on scales large compared with those of driving flow or velocity field, and is formulated in terms of large-scale or mean component of the field. See [51, 79] for discussion of the validity of mean field dynamo equations. The crucial assumption in these theories is an enhanced turbulent diffusivity $\sim W\ell$ which is responsible for dissipating the small-scale magnetic fields and transporting the residual large-scale poloidal fields downward to the tachocline region where the differential rotation produces the toroidal field. This scenario has been questioned by Piddington [88] on the ground that the buoyant weak poloidal field is unlikely to sink to the tachocline in which the toroidal field is generated. Note, the α -effect and Ω-effect occur in different layers in Parker's dynamo model, while Steenbeck and Krause [110] describe a dynamo model in which in the Ω-effect generating the toroidal field by differential rotation and the α-effect restoring the poloidal field from the toroidal component with the aid of helical turbulence, both occurring in the same layer.

It is noteworthy that Walen [127], Cowling [6], Piddington [88] appealed to the torsional oscillations of the original dipole-like solar magnetic field. This prompted Layzer [56] to explore the scenario of the Sun having an irregular large-scale magnetic field left over from the Hayashi phase that is confined to uniformly rotating radiative core. Such a seed field could be generated by the Bierman's [16] "battery" mechanism which is capable of producing large-scale toroidal magnetic field in ionized medium which grows at the rate $\propto \Omega^2$ which in turn is roughly comparable to the mean density.

In summary, it is now generally agreed that the toroidal field of the Sun is generated by differential rotation acting on a weak large-scale poloidal magnetic field. In the mean field dynamo theories the poloidal field is regenerated from the toroidal field through the process in which helical (cyclonic) turbulence, α-effect, and turbulent diffusivity, β play a major role, both are somewhat ad hoc inputs. On the other hand, the hydromagnetic oscillatory theory attributes reversal of the submerged toroidal magnetic field to the phenomenon of hydromagnetic torsional oscillations operating in the solar convection zone.

Thus, it seems that for driving the solar activity cycle, it is useful to invoke some sort of an oscillatory theory in which the kinetic energy of rotation is converted into magnetic energy in a cyclical manner with significant dissipation via turbulent convection and removal of the magnetic flux from the surface in order to avoid saturation of the dynamo action. Interestingly, the helioseismic observation indicate the kinetic energy of rotation declining in the equatorial latitudes from the solar minimum to the maximum phase when integrated over bulk of the convection zone. On the other hand, the residual temporally varying kinetic energy of rotation in the polar latitudes show an increasing trend from the minimum to the maximum phase of the activity cycle [1]. In other words, seismic observations suggest polar regions ($\geq$45 degree latitude) to speed up, while the equatorial regions to slow down with progress of the activity cycle, except that in the top 5% of the solar radius beneath the photosphere the differential rotational kinetic energy variation is in phase with the solar cycle. Should the kinetic energy of rotation be out of phase with the magnetic energy, this behaviour would support the build up of the magnetic activity near the maximum phase of the activity cycle, as observed. It is tempting to infer that angular momentum is being transported from the equatorial to polar latitudes as the solar cycle progresses, presumably by an agent such as meridional circulation in the new reversed poloidal field for the next cycle. Parker added the interaction of helical turbulent convection with the azimuthal component of the magnetic field to suggest that the Coriolis force on the rising and falling matter would cause the twisting of the toroidal field into the meridional so as to reinforce the poloidal magnetic field components.

(c) *Oblique rotator theory*

In this scenario, the star is supposed to rotate about an axis inclined at a substantial angle to the sight-line, its magnetic axis being also inclined at an angle to the rotation axis. Such a configuration could bring regions of positive and negative magnetic polarities alternately into view, thus leading to reversal of the observed magnetic field. This conceptually simple model is attractive from theoretical viewpoint but it is inadequate as a complete explanation because of the observed irregular changes often superimposed on the magnetic cycles.

Dynamo theory, which is applicable both to Earth, Sun and stars, and, even to the galaxy and galaxy clusters, attributes the generation and maintenance of the cosmic magnetic fields over a long time scales ($\geq 10^9$–10^{10} years) to electric currents induced in the interior of cosmic bodies as a result of material motion across the magnetic field lines, in the form of a self-excited dynamo. Of course, for getting the complete solution of the dynamo problem, it would be necessary to solve the hydrodynamic and electrodynamic equations simultaneously, i.e. to show both the existence of such a motion capable of maintaining the observed magnetic field for a long duration, and that this motion may be sustained against the dissipation due to viscosity and magnetic diffusivity by the forces available for the system under consideration.

It is somewhat prohibitive to handle such a formidable problem and for this reason it is customary to restrict the analysis to the "kinematic" dynamo model in which the motion is regarded as given or assigned and the problem simply amounts to whether the motion is capable of maintaining a magnetic field against dissipation. For this purpose a number of different versions of the problem have been discussed:

1. Seek a steady motion which is capable of maintaining a steady or periodically varying magnetic field.
2. Suppose the motion and the magnetic field are temporally fluctuating, but nevertheless possessing certain permanent characteristics, e.g. turbulent motion with associated irregular (turbulent) small-scale magnetic fields.
3. Study of an intermediate case with a small-scale travelling magnetic field [22].

Currently, the literature is largely concentrated on studying the mean-field dynamo with its associated ad hoc parameters. A major development in the dynamo theory is Cowling's anti-dynamo theorem known for establishing in a two-dimensional geometry the impossibility of generating an axisymmetric magnetic field by the dynamo action of an axisymmetric velocity field. Cowling's anti-dynamo theorem states that a steady axially symmetric magnetic field cannot be maintained by axisymmetric motions in the presence of finite dissipation, neglecting the surface currents in the analysis. Recently, it has been demonstrated by Plunian and Albousiere [90] that adopting an anisotropic electrical conductivity (resistivity) it is, indeed, possible to operate an axisymmetric dynamo with even a rigid body rotation.

Consider the situation where the velocity and the magnetic field are both in the meridional plane, and the lines of force are closed curves as a result of div $\mathbf{B} = 0$. The magnetic field will then have two limiting neutral points, N in the meridional plane at which the magnetic field vanishes. Now, because the lines of force are closed curves round N, the magnetic field would vanish at the neutral points, but since by Ampere's law, curl $\mathbf{H} = 4\pi\mathbf{j}/c$, neglecting the displacement current, the current density $\mathbf{j}$ will not vanish. Adopting the cylindrical geometry (w, θ, z) where for axisymmetric case, we have $u_\theta = 0$ and $H_\theta = 0$, the current density has only the θ-component now, noting that the e.m.f. $\mathbf{u} \times \mathbf{B}/c$ has a component only in the θ-direction, from Ohm's law, $\mathbf{j}/\sigma = \mathbf{E} + \mathbf{u} \times \mathbf{B}/c$, the electric field $\mathbf{E}$ has only the θ-component. For a steady dynamo, curl $\mathbf{E} = -\frac{1}{c}\partial\mathbf{B}/\partial t \equiv 0$, and consequently, integrating along small circle around the neutral points (Fig. 13.2)

$$\int \text{curl } \mathbf{E}.\hat{\mathbf{n}}dS = \oint \mathbf{E}.d\mathbf{r} \simeq E_\theta r d\theta = 0 \qquad (13.18)$$

which implies $E_\theta = 0$ and hence $\mathbf{E} \equiv 0$. At the neutral points, N, magnetic field $\mathbf{H}$ (or $\mathbf{B}$) vanish which means $\mathbf{u} \times \mathbf{B} = 0$. There is thus no induced e.m.f. by motion as well as no e.m.f. due to charges. But $\mathbf{j} \neq 0$ at the neutral points, N, which is impossible. This establishes Cowling's anti-dynamo theorem which can be

Fig. 13.2 When the velocity and the magnetic field are both in the meridional plane, the lines of force are closed curves as a result of div $\mathbf{B} = 0$. The magnetic field will then have two limiting neutral points, N shown here

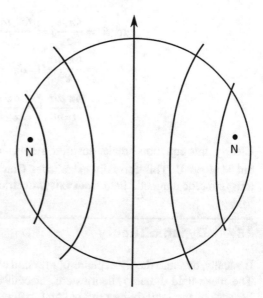

demonstrated alternatively by introducing the flux function ψ so that, in cylindrical geometry,

$$B_w = -\frac{1}{w}\frac{\partial \psi}{\partial z}, \qquad B_z = \frac{1}{w}\frac{\partial \psi}{\partial w}, \qquad (13.19)$$

satisfying div $\boldsymbol{B} = 0$. Then, the total flux through a circle with its centre on the axis is

$$\oint_0^w 2\pi w dw\, B_z = 2\pi \psi. \qquad (13.20)$$

Choose ψ to be zero on the axis ($w = 0$) and at infinity. There then exists a turning point such that $\frac{\partial \psi}{\partial w} = 0$, $\frac{\partial \psi}{\partial z} = 0$ but $\nabla^2 \psi \neq 0$. In a steady state, Ohm's law gives $\boldsymbol{j} = \sigma \boldsymbol{u} \times \boldsymbol{B}/c$. Note

$$\begin{aligned}
\text{curl}\,\boldsymbol{B} &= \frac{\partial}{\partial z}B_w - \frac{\partial}{\partial w}B_z \\
&= -\frac{1}{w}\frac{\partial^2 \psi}{\partial z^2} - \frac{1}{w}\frac{\partial^2 \psi}{\partial w^2} + \frac{1}{w^2}\frac{\partial \psi}{\partial w} \\
&= -\frac{1}{w}\nabla^2 \psi + \frac{1}{w^2}\frac{\partial \psi}{\partial w}. \qquad (13.21)
\end{aligned}$$

Using the Ampere's law,

$$\begin{aligned}
\text{curl}\,\boldsymbol{B} &= \frac{4\pi \mu}{c}\boldsymbol{j} = \frac{4\pi \mu \sigma}{c^2}(\boldsymbol{u} \times \boldsymbol{B}) \\
&= \frac{4\pi \mu \sigma}{c^2}(u_z B_w - u_w B_z) \\
&= \frac{4\pi \mu \sigma}{c^2 w}\left(u_z \frac{\partial \psi}{\partial z} + u_w \frac{\partial \psi}{\partial w}\right). \qquad (13.22)
\end{aligned}$$

Clearly, this equation breaks down at the turning point where $\frac{\partial \psi}{\partial w} = 0$, $\frac{\partial \psi}{\partial z} = 0$, but $\nabla^2 \psi \neq 0$. This formally establishes Cowling's anti-dynamo theorem for an axisymmetric magnetic field from axisymmetric motion.

13.4   Dynamo Theory

Basically, dynamo theory explains conversion of kinetic energy to magnetic energy. The mean-field dynamo theories are, of course, are not without their difficulties because of the underlying physical and mathematical assumptions. In fact, it was pointed out by Layzer et al. [57] that the dynamo equation is not a mathematically tenable or physically plausible approximation to the exact equations governing the mean-field, and consequently, there is no acceptable physical or mathematical basis for the crucial assumptions regarding the enhanced diffusion of weak magnetic field.

The principal physical mechanism underlying the solar activity cycle was introduced by Cowling [6] and later developed by Parker [86], Babcock [5], Leighton [61]. Cowling emphasized the first step in the dynamo theory of generating a strong submerged toroidal magnetic field by the action of observed solar differential rotation and resulting in the eruption of toroidal girdle due to the magnetic buoyancy to produce the sunspots at the solar surface. Parker went on to propose that the poloidal field itself is regenerated by an interaction between helical cyclonic turbulent convection and the rising toroidal girdle. Babcock [5] developed detailed solar cycle scenario incorporating the ideas of Cowling and Parker, and later, Leighton [60] proposed that the poloidal field will be regenerated from the diffusion and merging of the previous cycle sunspot fields transported by fluid motions which was further developed to reproduce the latitude-dependent time-varying "butterfly" diagram.

The observations with magnetograms revealed the photospheric magnetic field to be largely confined to flux tubes of $\sim$200 km diameter with $B \geq 10^3$ G field strength [66, 111]. And this raised doubts about the existence of large-scale poloidal magnetic field.

Further skepticism arose from Howard [40] observation of the solar surface rotation varying during to the course of the solar cycle, suggesting that the turbulent diffusion of magnetic field, a key ingredient in dynamo theories is probably not occurring in the sub-photospheric layers.

A major obstacle is the occurrence of Mounder minimum highlighted by Eddy [27] when the solar activity was nearly dormant for several decades. This observation is a major hurdle for standard dynamo models, since the dynamo action will be suppressed in the absence of poloidal magnetic field. The first step in the dynamo model is the conversion of the poloidal magnetic field into a toroidal component by the action of observed solar differential rotation. But the regenerating process of the poloidal field with the aid of helical (cyclonic) turbulence (kinetic helicity), in the presence of rotation seems rather contrived. In any event such an α-effect is likely to be quenched by the effect of a significantly large ($\geq 10^3$ G) magnetic field.

A major feature associated with the solar activity cycle is the "reversal" of the magnetic field polarity in the successive cycles. Clearly, in the process symbolically expressible as $\boldsymbol{B}_p \times \delta\Omega \rightarrow \boldsymbol{B}_t$ for generation of toroidal field $\boldsymbol{B}_t$ from the poloidal field with the aid of differential rotation, we must have either $\boldsymbol{B}_p$ changing sign every half-cycle or $\delta\Omega$ changing sign in the process of an oscillator phenomenon such as the observed torsional oscillations, the former in the framework of regenerative dynamo theories ($(\boldsymbol{u}_{\text{tub}} \times \boldsymbol{B}_t) \rightarrow \boldsymbol{B}_t$) and latter in the class of oscillator theories. Note the turbulent dynamo theories thus invoke convective turbulence to (a) generate small-scale fields from the large-scale toroidal field, (b) merge these small-scale fields to create a large-scale poloidal field comparable in strength to the original poloidal field, (c) combine the newly created poloidal field with the original poloidal field, (d) turbulent diffusion.

13.5   Understanding the Black Hole Evaporation Using Classical Fluid Flow

A black hole is a very massive compact object, say $10M_\odot$ mass, $M_\odot$ being a solar mass, equivalent to 10^{31} kg confined in a small volume, say, 30 km diameter such that its escape velocity becomes comparable to or greater than the speed of light. According to the General Theory of Relativity (GTR), gravity is a manifestation of the space-time curvature [126]. Massive compact objects like the black holes distort the space-time around them, creating a surface around the object called as "Event Horizon" through which nothing can come out, not even light.

Bekenstein [10] proposed that black holes should have a non-zero temperature and entropy. The entropy is proportional to the surface area of the event horizon of the black hole. In 1974, Hawking [39] proposed that the black holes are not completely black but emit very low energy thermal radiation with a black body spectrum equivalent to a Black Body Temperature $hc^3/8\pi k_B G M_{BH}$, where h is the Planck's constant, c is the speed of light, k_B is Boltzmann constant, G is Universal gravitational constant, M_{BH} is the mass of the black hole. This phenomenon is called as *black hole evaporation*. The mechanism of the evaporation of the black hole may be understood by quantum field theoretic approach in a curved space-time manifold. One of the consequences of the uncertainty principle of quantum mechanics is that it is possible for the law of conservation of energy to be violated locally for a very short period. The universe may produce mass and energy out of nowhere that may disappear very quickly—by a strange phenomenon called as *vacuum fluctuation*. Pairs of particles and antiparticles may appear and annihilate within a very short time. Now, it may happen that vacuum fluctuations occur very close to the event horizon at the exterior side of the black hole and one of the created two particles falls inside the black hole while the other escaped. The particle that escaped would carry energy away from the black hole and may be detected by a distant observer. If this process happens randomly, repeatedly and continuously, then the observer sees a continuous stream of radiation from the black hole.

During this Hawking radiation process, the lower the mass of a black hole, the higher the energy of the emitted radiation. As a black hole radiates, its mass reduces and it starts emitting more radiation causing it to evaporate more and more rapidly. Eventually, it shrinks to a size around the Planck's mass, the point at which its Compton wavelength equals to the Schwarzschild radius. At this point the dynamics of the black hole is expected to be governed by quantum gravity. The physics of the black hole evaporation works under the assumption that the quantum fields do not affect the gravitational field in which they propagate, the gravitational field itself is not quantized, and the wave equation for the quantum field is valid at all scales.

One may estimate the amount of total power radiated by a non-rotating Schwarzschild black hole [78]. The Schwazschild radius is given by

$$r_S = \frac{2GM_{BH}}{c^2}. \tag{13.23}$$

It gives the Schwarzschild surface area:

$$A_S = 4\pi r_S^2 = \frac{16\pi G M_{BH}}{c^4},$$ (13.24)

which provides the total power radiated at temperature T for emissivity ϵ as

$$P_T = A_S \epsilon \sigma T^4 = \frac{\hbar c^6}{15,360\pi G^2 M_{BH}^2},$$ (13.25)

where emissivity for the black body $\epsilon = 1$, Stefan's constant $\sigma = \frac{\pi^2 k_B^4}{60\hbar^3 c^2}$. For a $10 M_\odot$ (2×10^{31} kg) black hole, total radiated power could be approximated to be around 8.91×10^{-49} W. So

$$-\frac{dM_{BH}}{dt} = \frac{dE}{dt} = A_S \sigma T^4.$$ (13.26)

Now let us assume that black hole having initial mass M evaporates such that after a time t_{evap}, nothing remains. Hence, (13.26) reduces to

$$\int_0^{t_{evap}} dt = - \int_{M_{BH}}^0 \frac{1}{A\sigma T^4} dm, \quad \text{or,} \quad t_{evap} = \frac{M_{BH}}{A\sigma T^4},$$

which yields, $t_{evap} = (2.095 \times 10^{67}) \left(\frac{M_{BH}}{M_\odot}\right)^3$ years. (13.27)

This is much more than the age of the Universe, estimated about 14 billion years.

In 1981, Unruh [118, 119], provided a theoretical prescription to show that a supersonic classical fluid flow may give rise to a phenomenon analogous to black hole evaporation. The idea is that a moving fluid would drag sound waves along with it. If the speed of the fluid ever becomes supersonic, then inside this supersonic fluid, sound waves will never be able to return upstream. Keeping the analogy with GTR, kinematic features of fluid are to be used to determine the nature of field distributions in curved space-time, thereby generating an effective space-time metric. The analogy could be made at two levels: (a) geometrical acoustics, (b) physical acoustics. The geometric acoustics method is relatively simpler to deduce causal structure of space-time where unique effective metric could not be obtained.

13.5.1 Fluid Dynamics and the Geometric Metric for Massless Scalar Field

Let us consider that matter around the black hole is entering such that the fluid approximation may be considered with respect to the matter density in the accretion disk of the black hole. Let us consider an irrotational fluid flow as

$$\frac{\partial \rho}{\partial t} + \nabla \cdot (\rho \mathbf{v}) = 0, \tag{13.28}$$

$$\frac{\partial \mathbf{v}}{\partial t} + (\mathbf{v} \cdot \nabla) \mathbf{v} = -\frac{\nabla p}{\rho} - \nabla \Phi, \tag{13.29}$$

$$\nabla \times \mathbf{v} = 0, \tag{13.30}$$

where, ρ is the density of the fluid, $\mathbf{v}$ is the fluid speed, p is the fluid pressure due to hydrodynamic flow and is assumed to be the function of the density and Φ is an external force potential. The irrotational fluid flow $\nabla \times \mathbf{v}$ allows us to construct a velocity potential ψ such that

$$\mathbf{v} = \nabla \psi. \tag{13.31}$$

To define the sound speed inside the fluid, let us define a new function $g : g = g(\rho)$ such that

$$c^2(\rho) = \frac{\partial g (\ln |\rho|)}{\partial \ln |\rho|} = \frac{\partial p}{\partial \rho},$$

$$\Rightarrow \quad g (\ln |\rho|) = g(\xi) = \int^\rho \frac{dp}{d\rho'} d(\ln |\rho'|),$$

$$\Rightarrow \quad \xi = \ln |\rho|. \tag{13.32}$$

$$\Rightarrow \quad \frac{\partial \xi}{\partial t} = \frac{1}{\rho} \nabla \rho \text{ and } \nabla \xi = \frac{1}{\rho} \nabla \rho. \tag{13.33}$$

Substituting Eq. (13.31) into (13.29), we get

$$\nabla \left(\frac{\partial \psi}{\partial t} \right) + \frac{1}{2} (\mathbf{v} \cdot \mathbf{v}) + g(\xi) + \Phi = 0. \tag{13.34}$$

Integrating, we get

$$\frac{\partial \psi}{\partial t} + \frac{1}{2} (\mathbf{v} \cdot \mathbf{v}) + g(\xi) + \Phi = 0. \tag{13.35}$$

Substituting Eq. (13.33) into (13.28), we get

$$\frac{\partial \xi}{\partial t} + \mathbf{v} \cdot \nabla \xi + \nabla \cdot \mathbf{v} = 0. \tag{13.36}$$

Let us linearise (13.35) and (13.36) around some steady background ψ_0, ρ_0 and p_0 so that

$$\rho = \rho_0 + \epsilon \rho_1 + O(\epsilon^2), \tag{13.37}$$

$$p = p_0 + \epsilon p_1 + O(\epsilon^2), \tag{13.38}$$

$$\psi = \psi_0 + \epsilon \psi_1 + O(\epsilon^2), \tag{13.39}$$

which implies

$$\psi = \psi_0 + \tilde{\psi}, \tag{13.40}$$

$$\xi = \xi_0 + \tilde{\xi}. \tag{13.41}$$

Equation (13.35) reduces to

$$\frac{\partial}{\partial t} \left(\psi_0 + \tilde{\psi} \right) + \frac{1}{2} \left(\mathbf{v_0} + \tilde{\mathbf{v}} \right) \cdot \left(\mathbf{v_0} + \tilde{\mathbf{v}} \right) + g \left(\xi_0 + \tilde{\xi} \right) + \Psi = 0,$$

$$\text{or,} \quad \frac{\partial \tilde{\psi}}{\partial t} + \left(\mathbf{v_0} \cdot \nabla \tilde{\psi} \right) + \tilde{\xi} g' \left(\xi_0 \right) = 0,$$

$$\text{or,} \quad \tilde{\xi} = - \left(\frac{\partial \tilde{\psi}}{\partial t} + \mathbf{v_0} \cdot \nabla \tilde{\psi} \right) \cdot \frac{1}{g' \left(\xi_0 \right)}. \tag{13.42}$$

Similarly, one may linearize Eq. (13.36) as

$$\frac{1}{\rho_0} \left[\frac{\partial}{\partial t} \left(\rho_0 \tilde{\xi} \right) + \nabla \cdot \left(\rho_0 \mathbf{v_0} \tilde{\xi} \right) \right] + \frac{1}{\rho_0} \nabla \cdot \left(\rho_0 \nabla \tilde{\psi} \right) = 0. \tag{13.43}$$

Inserting (13.42) we get,

$$\frac{1}{\rho_0} \left[\frac{\partial}{\partial t} \left\{ - \frac{\rho_0}{\frac{\partial g(\xi_0)}{\partial \xi}} \left(\frac{\partial \tilde{\psi}}{\partial t} + \mathbf{v_0} \cdot \nabla \tilde{\psi} \right) \right\} \right] + \frac{1}{\rho_0} \left[\nabla \cdot \left\{ - \frac{\rho_0 \mathbf{v_0}}{\frac{\partial g(\xi_0)}{\partial \xi}} \left(\frac{\partial \tilde{\psi}}{\partial t} + \mathbf{v_0} \cdot \nabla \tilde{\psi} \right) \right\} \right]$$

$$+ \frac{1}{\rho_0} \nabla \cdot \left(\rho_0 \nabla \tilde{\psi} \right) = 0. \tag{13.44}$$

But, $\frac{\partial g(\xi_0)}{\partial \xi} = c^2(\rho_0) = c^2$. So,

$$\frac{1}{\rho_0}\left[\frac{\partial}{\partial t}\left\{\frac{\rho_0}{c^2}\frac{\partial \tilde{\psi}}{\partial t}\right\} + \frac{\partial}{\partial t}\left\{\frac{\rho_0}{c^2}\left(\mathbf{v_0}\cdot\nabla\tilde{\psi}\right)\right\}\right.$$

$$\left. +\nabla\cdot\left\{\frac{\rho_0}{c^2}\mathbf{v_0}\frac{\partial \tilde{\psi}}{\partial t} + \frac{\rho_0}{c^2}\mathbf{v_0}\left(\mathbf{v_0}\cdot\nabla\tilde{\psi}\right)\right\} - \nabla\cdot\left(\rho_0\nabla\tilde{\psi}\right)\right] = 0, \qquad (13.45)$$

or,

$$\frac{\rho_0}{c^2}\left[\frac{\partial^2 \tilde{\psi}}{\partial t^2} + \frac{\partial}{\partial t}\left(\mathbf{v_0}\cdot\nabla\tilde{\psi}\right) - c^2\left(\nabla\cdot\nabla\tilde{\psi}\right) + \nabla\cdot\left(\mathbf{v_0}\frac{\partial \tilde{\psi}}{\partial t}\right) + \nabla\cdot\left\{\mathbf{v_0}\left(\mathbf{v_0}\cdot\nabla\tilde{\psi}\right)\right\}\right] = 0.$$

$$(13.46)$$

That is,

$$\frac{\rho_0}{c^2}\left[-\left(c^2 - \mathbf{v_0}\cdot\mathbf{v_0}\right)\left(\nabla\cdot\nabla\tilde{\psi}\right) + 2\mathbf{v_0}\cdot\frac{\partial}{\partial t}\left(\nabla\tilde{\psi}\right) + \frac{\partial^2 \tilde{\psi}}{\partial t^2}\right] = 0. \qquad (13.47)$$

Integrating twice with respect to t

$$\frac{\rho_0}{c^2}\left[-\left(c^2 - \mathbf{v_0}\cdot\mathbf{v_0}\right)\left(\nabla\cdot\nabla\tilde{\psi}\right)dt^2 + 2dt\mathbf{v_0}\cdot\nabla\tilde{\psi} + \tilde{\psi}\right] = 0. \qquad (13.48)$$

Integrating twice with respect to r

$$\frac{\rho_0}{c^2}\left[-\left(c^2 - \mathbf{v_0}\cdot\mathbf{v_0}\right)\left(\nabla\cdot\nabla\tilde{\psi}\right)dt^2 + 2dt\mathbf{v_0}\cdot\mathbf{dr}\tilde{\psi} + \mathbf{dr}\cdot\mathbf{dr}\tilde{\psi}\right] = 0,$$

or $$\frac{\rho_0}{c^2}\left[-\left\{c^2 - \mathbf{v_0}\cdot\mathbf{v_0}\right\} + 2\mathbf{v_0}\cdot\frac{\mathbf{dr}}{dt} + \frac{\mathbf{dr}}{dt}\cdot\frac{\mathbf{dr}}{dt}\right] = 0. \qquad (13.49)$$

The wave equation (13.45) completely determines the propagation of the acoustic disturbances. Equation (13.47) may be expressed as

$$\partial_\mu\left[f^{\mu\nu}\partial_\nu\tilde{\psi}\right], \qquad (13.50)$$

where, $f^{\mu\nu}$ is suitable 4×4 symmetric matrix constructed explicitly from (13.47) such that μ, ν go from 0 to 3 where, $x^0 \equiv t$, $x^1 \equiv x$, $x^2 \equiv y$ and $x^3 \equiv z$, that is

$$f^{\mu\nu}(t, \mathbf{r}) \equiv \frac{\rho_0}{c^2}\begin{bmatrix} -1 & \vdots & -v_0^j \\ \cdots & \cdots & \cdots \\ -v_0^i & \vdots & c^2\delta^{ij} - v_0^i v_0^j \end{bmatrix}, \qquad (13.51)$$

where, δ^{ij} is Kronecker's delta function and all the indices run from 0 to 3. Thus expanding (13.51)

$$\text{Det}\left[f^{\mu\nu}\right] = \left(\frac{\rho_0}{c^2}\right)^4 \cdot (-c^6) = -\frac{\rho_0^4}{c^2}. \tag{13.52}$$

In any Lorentzian, that is, pseudo-Riemannian manifold, the curved space scalar d'Alembertian is expressed in terms of the metric $g_{\mu\nu}(t, \mathbf{r})$ as

$$\Box = \partial_\mu \partial^\mu = g_{\mu\nu}\partial^\mu\partial^\mu = \frac{\partial^2}{\partial t^2} - \nabla^2 = \frac{\partial^2}{\partial t^2} - \Delta, \tag{13.53}$$

where, usual $c = 1$ is chosen. Therefore, using the operator in (13.53) one may write

$$\Delta\tilde{\psi} \equiv \frac{1}{\sqrt{-g}}\partial_\mu\left[\sqrt{-g}g^{\mu\nu}\partial_\nu\tilde{\psi}\right], \tag{13.54}$$

where, $g^{\mu\nu}(t, \mathbf{r})$ is pointwise matrix inverse of $g_{\mu\nu}$ while $g \equiv \det(g_{\mu\nu})$. From (13.50) and (13.54),

$$\sqrt{-g}g^{\mu\nu} = f^{\mu\nu}$$
$$\Rightarrow det(f^{\mu\nu}) = \left(\sqrt{-g}\right)^4 g^{-1} = g. \tag{13.55}$$

Putting (13.52) in (13.55),

$$g = -\frac{\rho_0^4}{c^2} \text{ and } \sqrt{-g} = \frac{\rho_0^2}{c}. \tag{13.56}$$

Therefore, the acoustic line interval may be written as,

$$\begin{aligned} ds^2 &= g_{\mu\nu}dx^\mu dx^\nu, \\ &= \frac{\rho_0}{c}\left[-c^2dt^2 + \left(dx^i - v_0^i dt\right)\delta_{ij}\left(dx^j - v_0^j dt\right)\right], \\ &= \frac{\rho_0}{c}\left[\left\{c^2 - \mathbf{v_0} \cdot \mathbf{v_0}\right\}dt^2 + 2dt\mathbf{v_0} \cdot \mathbf{dr} - \mathbf{dr} \cdot \mathbf{dr}\right], \end{aligned} \tag{13.57}$$

It is important to note that two distinct metrics are involved in the system. One is the usual flat metric in Minkowski space involving physical space-time metric as

$$\eta_{\mu\nu} \equiv \left(\text{diag}\left[-c^2, 1, 1, 1\right]\right), \tag{13.58}$$

where c is the speed of light. Fluid elements actually couple to $\eta_{\mu\nu}$. Fluid motion is non-relativistic and Galilean invariance is valid.

Sound waves do not feel this physical metric at all. Any acoustic perturbation that may arise couple to acoustic metric only. We consider integral curves of vectors as

$$K^\mu \equiv \left(\frac{\partial}{\partial t}\right)^\mu = (1, 0, 0, 0)^\mu .$$ (13.59)

In steady flow, if K^μ preserves metric then it is called as *Killing vector*. In that case,

$$g_{\mu\nu}\left(\frac{\partial}{\partial t}\right)^\mu \left(\frac{\partial}{\partial t}\right)^\nu = g_{tt} = -\left[c^2 - v_0^2\right],$$ (13.60)

which changes sign if $|v_0| > c$. This generates supersonic flow to create ergo-region. In GTR, ergo sphere surrounding any spinning black hole is a region of space which moves with superluminal velocity with respect to the fixed stars.

If any set of points is displaced by $X^i dX_i$ where all distance relationship remain unchanged, that is, existence of isometry, then the vector field is called Killing vector.

At this point it is important to understand idea of trapped surface. Consider any closed two-surfaces. If the fluid velocity everywhere is pointing inward and the normal component of the fluid velocity is everywhere greater than the local speed of sound wave, it will be swept inward by the fluid flow and be trapped inside the surface. The surface is then called as "Outer Trapped Surface". For "Inner Trapped Surface" fluid flow is pointing everywhere outward.

Considering (13.57), if the vector $\mathbf{v_0}/(c^2 - v_0^2)$ is integrable, then one may define new time as

$$d\tau = dt + \frac{\mathbf{v_0} \cdot \mathbf{dr}}{c^2 - v_0^2},$$ (13.61)

$$\text{or, } dt = d\tau - \frac{\mathbf{v_0} \cdot \mathbf{dr}}{c^2 - v_0^2},$$

$$\text{or, } dt^2 = d\tau^2 + \frac{(\mathbf{v_0} \cdot \mathbf{v_0})\,(\mathbf{dr} \cdot \mathbf{dr})}{c^2 - v_0^2} - 2d\tau\frac{\mathbf{v_0} \cdot \mathbf{dr}}{c^2 - v_0^2}.$$ (13.62)

Substituting this into (13.57), we get

$$ds^2 = \frac{\rho_0}{c}\left[\left(c^2 - \mathbf{v_0} \cdot \mathbf{v_0}\right)d\tau^2 - \frac{(\mathbf{v_0} \cdot \mathbf{v_0})\,(\mathbf{dr} \cdot \mathbf{dr})}{c^2 - v_0^2} - \mathbf{dr} \cdot \mathbf{dr}\right].$$ (13.63)

It is important to observe that the absence of space-time cross term indicates the acoustic geometric is static. The condition that the acoustic geometry is static rather than stationary is seen to be

$$\nabla \times \left(\frac{\mathbf{v_0}}{c^2 - v_0^2} \right) = 0. \tag{13.64}$$

Since the fluid is irrotational, $\nabla \times \mathbf{v_0} = 0$. This implies

$$\mathbf{v_0} \times \nabla \left(c^2 - v_0^2 \right) = 0. \tag{13.65}$$

At this point, one may think about the concept of *event horizon* for a sound wave. Let us assume that at some arbitrary value $r = R$, fluid flow becomes supersonic as fluid is moving inwards in a convergent flow. Then one may express the speed v_0 as

$$v_0 = -c + \alpha(r - R). \tag{13.66}$$

Neglecting the second order terms,

$$v_0^2 \approx c^2 - 2c\alpha(r - R). \tag{13.67}$$

Putting (13.67) in (13.63)

$$ds^2 = \frac{\rho_0}{c} \left[2c\alpha(r - R)d\tau^2 - \frac{cdr^2}{2\alpha(r - R)} \right], \tag{13.68}$$

where $\mathbf{dr} \cdot \mathbf{dr}$ in spherically symmetric geometry becomes $r^2(d\theta^2 + \sin^2\theta d\phi^2)$. This can be neglected as we are interested in radial effects only. We have also neglected terms of order $O((r - R)^2)$.

One may set up a fiducial observer by properly normalizing the Killing vector as $\bar{K}/||\bar{K}||$. One may define the four-acceleration as

$$\bar{A} = \left(\frac{\bar{K}}{||\bar{K}||} \cdot \nabla \right) \frac{\bar{K}}{||\bar{K}||} \sim \frac{1}{2} \frac{\nabla ||K||^2}{||K||^2}. \tag{13.69}$$

Since the surface gravity is defined as $||\bar{A}|| ||K||$, one may write

$$
\begin{aligned}
K_S &= \frac{1}{2} \frac{\nabla ||K||^2}{||K||^2} \cdot ||K|| = \frac{1}{2} \frac{\nabla ||K||^2}{||K||}, \\
&= \frac{1}{2} \sqrt{\frac{c}{\rho_0}} \frac{1}{\sqrt{c^2 - v_0^2}} \frac{\partial}{\partial x} \left[\frac{\rho_0}{c} (c^2 - v_0^2) \right], \\
&= v_0 \sqrt{\frac{\rho_0}{c(c^2 - v_0^2)}} \frac{\partial v_0}{\partial x}.
\end{aligned}
\tag{13.70}
$$

At the event horizon

$$\left.\frac{\partial v_0}{\partial x}\right|_{\text{horizon}} \sim \frac{c}{R}. \tag{13.71}$$

Equations (13.70) and (13.71) gives the Hawking radiation temperature as

$$T_H = \frac{\hbar}{2\pi k_B}\frac{c}{R}, \tag{13.72}$$

where R is the event horizon.

An Introduction to Classical Turbulence

> Here it is difficult as it were to keep our heads up, - to see that
> we must stick to the subjects of our everyday thinking, and not
> go astray and imagine that we have to describe extreme
> subtleties, which in turn we are after all quite unable to describe
> with the means at our disposal. We feel as if we had to repair a
> torn spider's web with our fingers.
>
> —Ludwig Wittgenstein

Macroscopic classical physics poses challenges in research which remain modern even when the governing equations are from the times of Huygens and Newton. Turbulence is distinct from chaos even when several insights are shared. The monumental work carried out over several centuries to address many aspects of turbulent flow has given birth to ideas and tools like negative temperature, anomalous diffusion, power-law scaling in many-body problems, scale invariance, universality and so on.

For incompressible flow, the fundamental equations for velocity and pressure field are the Navier-Stokes equation and that the velocity field is solenoidal. To quantify this, using a typical velocity, U and a length scale, L, a dimensionless measure of nonlinearity, Reynolds number, is defined as $R = UL/\nu$ where ν is the kinematic molecular viscosity. For typical fluids we come across in real life, $\nu \sim 0.01$–0.1 cm^2/s. With typical velocities much larger than 1 cm/s and length scales much larger than 1 cm, $R \gg 100$; indeed, in a typical geophysical situation, $R \sim 10^8$. Nineteenth century brought the classification of the fluid flow broadly as laminar and turbulent for respectively low and high values of R. This chapter is about high Reynolds number flows where the difficulty arises due to appearance of many scales of motion. The largest scales correspond to turbulent energy and the smallest scales are where dissipation into heat takes place. The complexity of turbulent flow can be gauged by estimating the number of active degrees of freedom which varies as $R^{9/4}$ per unit volume—this gives a staggering figure of 10^{18}. The

© The Author(s), under exclusive license to Springer Nature Switzerland AG 2022   217
S. R. Jain et al., *A Primer on Fluid Mechanics with Applications*,
https://doi.org/10.1007/978-3-031-20487-6_14

excitations in turbulent flows does not allow any clear separation of scales, thus challenging a "solution" to the problem of turbulence. There are four basic concepts [34] which constitute the basis of our understanding of turbulence. We discuss these now.

Randomness

Due to the complex and unpredictable character of turbulent flows, statistical theory was developed where one concentrates on certain average values of some quantities. A plausible mechanism by which turbulent flows become random is the amplification of fluctuations in forces and noise due to intrinsic instabilities. This corresponds to the phenomenon of chaos where the key idea is sensitivity to initial conditions. Following the pioneering work of Poincaré, according to Landau, the manner in which chaos emerges is by a series of bifurcations or instabilities. Ruelle and Takens showed that chaotic behaviour results after a finite number of bifurcations. Most of the illustrations of the theory of dynamical systems has been restricted to lower-dimensional systems. However, turbulent flows generally correspond to attractors with higher dimensions.

Eddy Viscosity

Drawing an analogy from statistical mechanics, the concept of eddy viscosity was introduced by Boussinesq and developed by Taylor and Prandtl. The relations of transport coefficients to the time correlation function of certain microscopic quantities are of great significance. The two dominant scales in a kinetic description of gases are mean free path and diffusive (hydrodynamic) scale. Transport coefficient like kinematic viscosity are of the order of velocity of thermal motion times the mean free path. Prandtl proposed that small-scale eddies could be thought to act diffusively on large-scale eddies, and this may be quantified in terms of eddy viscosity. Eddy viscosity, ν_{eddy} is of the order of the product of root mean square velocity v_{rms} and mixing length, ℓ. Gradients in the mean velocity are smoothened by ν_{eddy}. Interestingly, the ratio of ν_{eddy} to ν is approximately R. Thus, transport is increased for turbulent flows in comparison to laminar flows. This implies that a passive scalar will diffuse much faster in a turbulent flow than in a laminar flow. This is also accompanied by an increased kinetic energy dissipation. In an incompressible flow, the local rate of viscous dissipation of energy per unit mass is $\mathcal{E} = \nu |\nabla \boldsymbol{v}|^2$. An estimate of this requires estimates of $|\nabla \boldsymbol{v}|$, particularly at the smallest scales of turbulent motion. The latter is generally inaccessible due to practical circumstances and only mean motions are measured. Employing this record along with ν_{eddy} gives us a good estimate of average dissipation rate. Note that on the scale of eddying motion, $|\nabla \boldsymbol{v}| \sim O(v_{\text{rms}}/L)$. Dissipation at the scale, L occurs through $\nu_{\text{eddy}} \approx v_{\text{rms}} L$. Thus, $\mathcal{E} \approx v_{\text{rms}} L \left(\frac{v_{\text{rms}}}{L}\right)^2 \approx \frac{v_{\text{rms}}^3}{L}$. Observe that this turns out to be independent of ν—verified by experiments and computer simulations. Moreover, a mere re-writing reveals that

$$\mathcal{E} \approx \frac{v_{\text{rms}}^2}{(L/v_{\text{rms}})} = \frac{\text{Kinetic energy per unit mass}}{\text{time taken for an eddy of size, } L \text{ to flip over}}. \quad (14.1)$$

Cascade

Richardson [98] proposed a hierarchical model of turbulence involving eddies of varying sizes. Produced by forces driving the flow, the large eddies are unstable and hence, in turn, they produce smaller eddies. These also become unstable and produce even smaller eddies. This continues until kinematic viscosity suppresses further cascading. This picture has been influential in shaping our understanding.

Scaling

In addition to all the invariance principles and conservation laws of mechanics, the hydrodynamic equations admit a new symmetry which appears in the limit of infinite Reynolds number. This forms the basis of the scaling theory of Kolmogorov [50]. To begin with, it is easy to note that if we ignore viscosity, the incompressible Navier-Stokes equation is invariant if distance scales by λ, velocity by λ^h, and time by λ^{1-h} where h is an arbitrary scaling exponent. Let us now enlist first, the assumptions, and then, the consequences of Kolmogorov theory.

Assumptions

(A1)   Scale invariance mentioned above holds in the statistical sense; the average quantities are assumed to be scale-invariant and not the detailed structures.

(A2)   Finite flux of energy $\mathcal{E}$ is assumed to flow from large scales where turbulence is produced, to small scales where it is dissipated as $R \to \infty$.

(A3) $\mathcal{E}_\ell$ through a scale ℓ is assumed to depend on flow quantities local to scale ℓ—velocities v_ℓ of eddies of size ℓ. Because $\mathcal{E}_\ell$ has dimensions of energy per unit mass, by dimensional analysis, $\mathcal{E}_\ell \approx v_\ell^3/\ell = v_\ell^3 \lambda^{3h}/\ell\lambda$, so scales as λ^{3h-1}. Scale invariance of $\mathcal{E}$ implies that $h = 1/3$.

Consequences

(C1)   From above discussion, $v_\ell \approx \mathcal{E}_\ell^{1/3}\ell^{1/3}$, thus $|\nabla v_\ell| \approx \mathcal{E}^{1/3}\ell^{-2/3}$, which shows the dominance by small scales in a turbulent flow. To control the divergence, it must be cut off at some small scale [84].

(C2)   The structure function of order, defined as $(\delta v_\ell)^p$, is approximately $\mathcal{E}^{p/3}\ell^{p/3}$; one says that the scaling exponent is $\zeta_p = p/3$.

(C3)   The Fourier transform of the structure function of order two is related to the energy spectrum; this satisfies the Kolmogorov-Obukhov law:

$$E(k) = C_{KO}\, \mathcal{E}^{2/3}\, k^{-5/3} \quad (14.2)$$

where k is the wavenumber and C_{KO} is Kolmogorov-Obukhov constant. Alternatively, this relation can be physically interpreted in the following way: In the inertial range, constancy of energy cascade from one scale to the another implies that

$$\mathcal{E} \simeq v_\ell^3/\ell \sim v_0^3/l_d$$

In other words, energy cascade rate must be balanced by the viscous dissipation rate. This follows

$$v_\ell \sim (\mathcal{E}\ell)^{\ell/3}$$

Finally, since $\int E(k)dk \sim v_\ell^2$, the energy spectrum must satisfy

$$E(k) \sim v_\ell^2 \ell \sim (\mathcal{E}\ell)^{2/3}\,\ell \sim \mathcal{E}^{2/3} k^{-5/3}$$

(C4)   The eddy viscosity at scale ℓ scales as $v_\ell \approx \ell v_\ell \approx \mathcal{E}^{1/3}\ell^{4/3}$.

Validity of Scaling Theory

Scaling symmetry holds in a range of length scales which is much smaller than the scale L where turbulence is produced. This is the scale set by external forces and boundaries. At the smallest scales where dissipation into smallest eddies occurs, kinematic viscosity cannot be ignored. Thus the scale of applicability ℓ must be much larger than the Kolmogorov dissipation scale, $\ell_d \approx (v^3/\mathcal{E})^{1/4}$. Scaling theory is expected to work in the inertial range, $\ell_d \ll \ell \ll L$. In terms of R,

$$\frac{L}{\ell_d} \approx \frac{v_{\text{rms}}^{3/4} L^{3/4}}{v^{3/4}} = R^{3/4}. \tag{14.3}$$

In three-dimensional flow, there are $R^{9/4}$ active degrees of freedom per unit volume L^3.

Kolmogorov's theory has been experimentally verified several times, beginning with the work of Grant et al. [36] where $0.0001 < k\ell_d < 1$.

With this background, we would like to delve deeper into some of the most important problems. The selection is for the purpose of illustration.

14.1   Taylor-Couette Flow: From Newton to Chandrasekhar

Fluid flow between two concentrically placed, rotating cylinders has been a subject of interest since the time of Newton. The varied patterns and chaos that it exhibits has led to an endless investigation, unravelling novel insights by degrees. The flow of fluid between rotating cylinders is called the Taylor-Couette flow—a careful consideration of this was carried out by Maurice Couette (1890), followed by a mathematically rigorous treatment by Geoffrey Ingram Taylor [114]. The subject can be traced back to Isaac Newton (1687) who considered circular motion of fluids [82]. He introduced what is now considered as a definition of Newtonian fluid as one where viscous stresses are proportional to the velocity gradient. The fundamental importance of Taylor-Couette flow is underlined by Russell Donnelly by the remark that it is the "hydrogen atom of fluid dynamics".

If both cylinders are turned, Stokes [112] considered that they should turn oppositely. If only one cylinder moved, it should be the outer one. If the inner one is made to rotate too fast, fluid near it would be unstable, flying outwards due to centrifugal force and produce eddies. This was settled by an analysis by Taylor in 75 years. Stokes also noted that for relatively small angular velocities, the fluid motion is roughly planar, it is not important where the fluid is terminated. Stokes noted that the boundary conditions at the solid surfaces were unknown—and this took almost a hundred years until the no-slip condition was used in the treatment by Taylor. Stokes was also concerned with the boundary conditions of the partially filled cylinders. Remarkably, he suggested the use of tracer (dust particles) to mar the flow. In time, this idea led to the laser Doppler velocimeter which measures the fluid velocity by observing the Doppler shift of scattered light from tracer particles in the fluid.

Equations of motion for a viscous fluid were formulated by Navier (1823) and Stokes (1845). It was realized by the experimenters that there are broadly two types of motion—laminar and turbulent. As the simple integrals of motion did not exist for turbulent flow, a higher value of viscosity was expected.

Max Margules [73] invented what is called the rotating cylinder type viscometer. In the same vein, and independently, viscosity of water was determined by Mallock in a series of pioneering contributions from 1888 onwards. His meticulously designed apparatus is shown in Fig. 14.1. The outer cylinder E rotates and produces torque on the inner cylinder A, containing water. Another cylinder G surrounds E, the gap between them is filled with water. the temperature in the annulus is taken to be the mean of the two thermometer readings. Air is trapped in a region in the bottom of A so that the fluid torque is exerted only on the cylindrical wall of A. The cylinder K is stationary. Mercury placed between K and E is to produce end conditions with the same velocity distribution as in water being measured. A telescope T is placed to read the displacement of the calibrated disk attached to stem B. The diameter of the cylinders is 15–20 cm and their height is about 25 cm.

Mallock found that when the inner cylinder is rotating, the torque and angular velocity are not linearly related, and he wrongly concluded that the flow is unstable. At the lowest speed of 2 rpm, Taylor vortices are formed for the size of cylinders used. With rotating outer cylinder, he found the flow was stable for low rotation rates and unstable for high rates.

Rayleigh studied stability in the absence of viscosity. He showed that the flow is stable if the angular momentum per unit mass of the fluid increases monotonically outward [19]. This implies that the motion is unstable if only the inner cylinder is

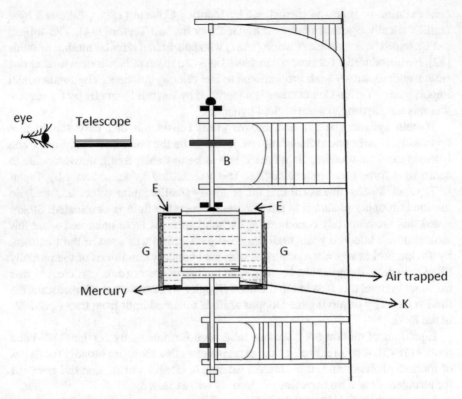

Fig. 14.1 Rotating cylinder type viscometer, designed by Max Margules (1888)

rotating; the motion is stable to infinitesimal perturbations if only the outer cylinder is rotating.

Maurice Couette, in his thesis (1890), concluded from his viscometer that there are two kinds of fluid motions—one given by exact integrals of motion and another at higher speeds where integrals are broken. He made a detailed study of viscosity using a pair of cylinders with the outer one rotating and the inner one suspended on a fibre to measure torque—today, these are called Couette viscometers.

Geoffrey Ingram Taylor [114] wrote one of most influential papers of the twentieth century on what is now called Taylor-Couette flow. The correspondence between theory and experiment for the stability found by him rested on the no-slip boundary condition for the flow at the solid surfaces. This success was taken as a proof of correctness of Navier-Stokes equation and if the no-slip condition for the fluid at the cylinder walls. Such use of Taylor-Couette flow to confirm fundamental ideas in fluid mechanics has become a tradition. For superfluid He II, the temperature-dependent onset of Taylor vortices is an instance, following similar tradition by Swanson and Donnelly [113]. These were predicted by Jones and Barenghi [45]. Chandrasekhar [19] generalized the Taylor-Couette flow to the case of conducting fluid in magnetic field.

14.1.1 Stability of Couette Flow

We present a brief treatment, following the classic work [19], on the stability of Couette flow—circular flow of a liquid between coaxial cylinders. This is a flow subjected to an adverse gradient of angular momentum. If Ω denotes the angular velocity of rotation of liquid about the axis, the general form is

$$\Omega(r) = A + \frac{B}{r^2} \tag{14.4}$$

in the absence of pressure gradient. The constants A and B depend on the angular velocities Ω_1, Ω_2 of inner and outer cylinders respectively. If R_1 and R_2 ($> R_1$) denote the radii of inner and outer cylinders, then

$$\Omega_1(r) = A + \frac{B}{R_1^2}, \qquad \Omega_2(r) = A + \frac{B}{R_2^2}. \tag{14.5}$$

Thus, with $\mu = \Omega_2/\Omega_1$, $\eta = R_1/R_2$, the constants are

$$A = -\Omega_1\eta^2\frac{1 - \mu/\eta^2}{1 - \eta^2}, \qquad B = \Omega_1\frac{R_1^2(1 - \mu)}{1 - \eta^2}. \tag{14.6}$$

The Navier-Stokes equations for incompressible fluids are

$$\frac{\partial u_r}{\partial t} + (\boldsymbol{u}.\nabla)u_r - \frac{u_\theta^2}{r} = -\frac{\partial}{\partial r}\left(\frac{p}{\rho}\right) + \nu\left(\nabla^2 u_r - \frac{2}{r^2}\frac{\partial u_\theta}{\partial \theta} - \frac{u_r}{r^2}\right), \tag{14.7}$$

$$\frac{\partial u_\theta}{\partial t} + (\boldsymbol{u}.\nabla)u_\theta + \frac{u_r u_\theta}{r} = -\frac{1}{r}\frac{\partial}{\partial \theta}\left(\frac{p}{\rho}\right) + \nu\left(\nabla^2 u_\theta + \frac{2}{r^2}\frac{\partial u_r}{\partial \theta} - \frac{u_\theta}{r^2}\right), \tag{14.8}$$

$$\frac{\partial u_z}{\partial t} + (\boldsymbol{u}.\nabla)u_z = -\frac{\partial}{\partial z}\left(\frac{p}{\rho}\right) + \nu\nabla^2 u_z \tag{14.9}$$

with

$$\nabla^2 = \frac{\partial^2}{\partial r^2} + \frac{1}{r}\frac{\partial}{\partial r} + \frac{1}{r^2}\frac{\partial^2}{\partial \theta^2} + \frac{\partial^2}{\partial z^2}. \tag{14.10}$$

The incompressibility condition is expressed as

$$\frac{\partial u_r}{\partial t} + \frac{u_r}{r} + \frac{1}{r}\frac{\partial u_\theta}{\partial \theta} + \frac{\partial u_z}{\partial z} = 0. \tag{14.11}$$

The stationary solutions $u_r = u_z = 0$, $u_\theta = V(r)$ exists if

$$\frac{d}{dr}\left(\frac{p}{\rho}\right) = \frac{V^2}{r} \tag{14.12}$$

and

$$\nu\left(\nabla^2 V - \frac{V}{r^2}\right) = \nu\frac{d}{dr}\left(\frac{d}{dr} + \frac{1}{r}\right)V = 0. \tag{14.13}$$

For $\nu \neq 0$, the most general solution for $V(r)$ is

$$V(r) = Ar + \frac{B}{r} \tag{14.14}$$

with corresponding angular velocity,

$$\Omega = A + \frac{B}{r^2} \tag{14.15}$$

as noted above. The stability for the viscous flow was first found theoretically and experimentally by Taylor. To study the stability, we write perturbed state as

$$u_r, V + u_\theta, u_z, \text{ and } \delta p/\rho = w \tag{14.16}$$

with axisymmetric perturbations. The linearized equations are

$$\frac{\partial u_r}{\partial t} - \frac{2V}{r}u_\theta = -\frac{\partial w}{\partial r} + \nu\left(\nabla^2 u_r - \frac{u_r}{r^2}\right), \tag{14.17}$$

$$\frac{\partial u_\theta}{\partial t} + \left(\frac{dV}{dr} + \frac{V}{r}\right)u_r = \nu\left(\nabla^2 u_\theta - \frac{u_\theta}{r^2}\right), \tag{14.18}$$

$$\frac{\partial u_z}{\partial t} = -\frac{\partial w}{\partial z} + \nu\nabla^2 u_z, \tag{14.19}$$

with $\nabla^2 = \partial^2/\partial r^2 + (1/r)\partial/\partial r + \partial^2/\partial z^2$; the condition of incompressibility is

$$\frac{\partial u_r}{\partial t} + \frac{u_r}{r} + \frac{\partial u_z}{\partial z} = 0. \tag{14.20}$$

Normal modes solutions may be written as

$$u_r = e^{pt}u(r)\cos kz, \qquad u_\theta = e^{pt}v(r)\cos kz,$$
$$u_z = e^{pt}w(r)\sin kz, \qquad w = e^{pt}w(r)\cos kz, \tag{14.21}$$

k being the wavenumber of the perturbation in the axial direction, and p a complex number. Following Chandrasekhar, denote d/dr $(d/dr + 1/r)$ by D (D_*), the linearized equations can be re-written as

$$v\left(D_*D - k^2 - \frac{1}{r^2} - \frac{p}{v}\right)u + 2\frac{V}{r}v = \frac{dw}{dr}, \tag{14.22}$$

$$v\left(D_*D - k^2 - \frac{1}{r^2} - \frac{p}{v}\right)v - (D_*V)u = 0, \tag{14.23}$$

$$v\left(D_*D - k^2 - \frac{p}{v}\right)w = -kw, \tag{14.24}$$

with

$$\nabla^2 = \left(\frac{d}{dr} + \frac{1}{r}\right)\frac{d}{dr} - k^2 = D_*D - k^2$$

$$= DD_* + \frac{1}{r^2} - k^2, \text{ and} \tag{14.25}$$

$$D_*u = -kw. \tag{14.26}$$

Eliminating w from these, we have

$$\frac{v}{k^2}\left(D_*D - k^2 - \frac{p}{v}\right)D_*u = w. \tag{14.27}$$

Inserting this w in (15.19), we find

$$\frac{v}{k^2}\left(DD_* - k^2 - \frac{p}{v}\right)(DD_* - k^2)u = 2\frac{V}{r}v. \tag{14.28}$$

Along with this, we have

$$v\left(DD_* - k^2 - \frac{p}{v}\right)v = (D_*V)u. \tag{14.29}$$

Defining $k^2 = a^2/R_2^2, \sigma = p\,R_2^2/v$, transforming $(2AR_2^2/v)u \to u$, we can re-write (15.25, 15.26) as

$$(DD_* - a^2 - \sigma)(DD_* - a^2)u = -Ta^2\left(\frac{1}{r^2} - \kappa\right)v, \tag{14.30}$$

$$(DD_* - a^2 - \sigma)v = u, \tag{14.31}$$

where

$$T = \frac{4\Omega_1^2 R_1^4}{v^2} \frac{(1-\mu)(1-\mu/\eta^2)}{(1-\eta^2)^2} \text{ (Taylor number)}, \quad \kappa = \frac{1-\mu/\eta^2)}{(1-\mu)}. \quad (14.32)$$

Equations (15.27) and (15.28) have to satisfy no-slip boundary conditions on the cylindrical walls at $r = 1, \eta$:

$$u = v = 0, \&, Du = 0 \text{ for } r = 1, \eta, \text{ or }, \text{equivalently } w = 0. \quad (14.33)$$

14.1.1.1 Rayleigh's Criterion
For an inviscid flow, Rayleigh showed that necessary and sufficient condition for a distribution, $\Omega(r)$ to be stable is

$$\frac{d}{dr}(r^2\Omega)^2 > 0 \quad (14.34)$$

everywhere. The flow is unstable if $(r^2\Omega)^2$ should decrease anywhere in the interval.

As we have taken $R_2 > R_1, \eta < 1$. For the invisicd case, the linearized equations lead to the stability condition, $\mu > \eta^2$. This is exactly the Rayleigh's criterion.

14.2   Chaos and Turbulence

Navier-Stokes equation is a nonlinear partial differential equation but it must be deterministic and the solutions must be predictable with well-defined initial conditions. The question we address here is whether there is a connection between well-known phenomenon of deterministic chaos and turbulence. Nonlinearity is common between the two. Let us try to set up what is called the "poor man's Navier-Stokes equation" by Frisch [33] for a discrete map:

$$v_{t+1} = 1 - 2v_t^2, \quad v_0 = w, \quad t = 0, 1, 2, \ldots \quad (14.35)$$

where v_t is real number between -1 and +1. Starting from an initial value, the iterates define an orbit. This is inspired from logistic map: $v \rightarrow v - av^2$. The map can be written in a way that it looks like Navier-Stokes equation:

$$v_{t+1} - v_t = -2v_t^2 - v_t + 1,$$
$$\frac{\partial \boldsymbol{v}}{\partial t} = -(\boldsymbol{v}.\nabla\boldsymbol{v} + \nabla p) + v\nabla^2\boldsymbol{v} + \mathbf{f}. \quad (14.36)$$

Substitute

$$v_t = \sin(\pi x_t - \pi/2), \quad v_{t+1} = \sin(\pi x_{t+1} - \pi/2), \quad (14.37)$$

$0 \le x_t \le 1$. This induces the tent map in x:

$$x_{t+1} = 2x_t, \quad 0 \le x_t \le 1/2,$$
$$= 2 - 2x_t, \quad 1/2 \le x_t \le 1. \tag{14.38}$$

In other words, logistic map is topologically conjugate to the tent map. The probability distribution function is given by

$$P(v)dv = \frac{1}{\pi} \frac{1}{\sqrt{1 - v^2}}. \tag{14.39}$$

The substitution above was first realized by von Neumann [125]. The logistic map exhibits a series of period-doubling bifurcations. The idea that bifurcation theory might be relevant to hydrodynamic stability was first suggested by Hopf (1942). Landau (1944) argued that bifurcations could play a role in understanding the transition to turbulence. Let us sketch the argument due to Landau. Let R be a control parameter in a hydrodynamic exponent. At a critical value of this parameter, R_c, there will be a transition to turbulence. For example, in an experiment where the liquid flows between concentric cylinders (Fig. 14.2), as the rotation rate of the inner

Fig. 14.2 Liquid flows between rotating concentric cylinders. As the rotation rate of inner cylinder increases, the flow becomes complex. Beyond a certain rate, Taylor vortices are formed

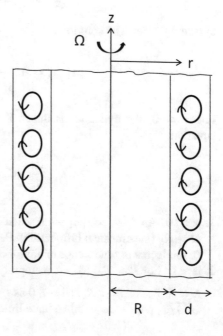

cylinder increases, the flow becomes progressively complex. At R_c, Taylor vortices are formed:

$$R_c = \sqrt{\frac{\Omega^2 d^3 \rho}{\nu^2}} = \sqrt{Ta} \tag{14.40}$$

where Ta is Taylor's number. At $R = R_c$, the amplitude and growth rate of the unstable normal mode is $A(t)$ and σ respectively:

$$A(t) = A_0 e^{(\sigma + i\omega)t}. \tag{14.41}$$

We have $\sigma < 0$ where $R < R_c$, and, $\sigma > 0$ for $R > R_c$ to say that about R_c the stability changes. We can write

$$\sigma \propto (R - R_c) + O[(R - R_c)^2], \quad |R - R_c| \ll R_c. \tag{14.42}$$

With increase in stability, Landau proposed that magnitude of A, averaged over many cycles, grows according to

$$\frac{d}{dt}|A|^2 = 2\sigma |A|^2 - \alpha |A|^4 \tag{14.43}$$

to leading order. The solution is

$$|A|^2(t) = A_0^2 \frac{e^{2\sigma t}}{1 + \frac{\alpha A_0^2}{2\sigma}(e^{2\sigma t} - 1)}. \tag{14.44}$$

For $\alpha = 0$, the dynamics is linear. For $\alpha \neq 0$, (14.43) resembles the logistic equation,

$$\frac{dy}{dt} = 2\sigma y - \alpha y^2. \tag{14.45}$$

1. $\alpha > 0$: As $t \to \infty$, $|A| \to 0$ when $R < R_c$, and, $|A| \to \sqrt{2\sigma/\alpha} = A_\infty$ for $R > R_c$ (supercritical bifurcation). Because $\sigma \sim (R - R_c)$, $A_\infty \sim (R - R_c)^{1/2}$ (reminiscent of emergence of Taylor cells).
2. $\alpha < 0$: For $R > R_c$, $|A| \to \infty$ rapidly within $t = t^* = \frac{1}{2\sigma} \log[1 + |\lambda|^{-1}]$ where $\lambda = \frac{\alpha A_0^2}{2\sigma}$. For $R < R_c$, $|A| \to 0$ as $t \to \infty$ if $A_0 < \sqrt{2\sigma/\alpha}$. If, however, $A_0 > \sqrt{2\sigma/\alpha}$, $|A| \to \infty$ within a finite time. This is called a subcritical bifurcation.

14.2.1 Lorenz Equations

Consider fluid confined in a narrow annular channel between stationary, horizontal cylindrical walls kept at a specified temperature $T_w(\phi)$ where ϕ is the polar angle. The fluid is heated from below. The fundamental equations are, of course,

$$\rho \frac{d\boldsymbol{v}}{dt} = \rho \left[\frac{\partial \boldsymbol{v}}{\partial t} + (\boldsymbol{v}.\nabla)\boldsymbol{v} \right] = -\nabla p + \rho \mathbf{g} + \eta \nabla^2 \boldsymbol{v},$$

$$\frac{dT}{dt} = \frac{\partial T}{\partial t} + \boldsymbol{v}.\nabla T = \kappa \nabla^2 T \qquad (14.46)$$

where $\mathbf{g} = -g\,\hat{z}$. The velocity of the viscous fluid flow along channel of width d is

$$\boldsymbol{v} = \frac{v_0}{(d/2)^2} [(d/2)^2 - x^2]\,\hat{z},$$

$$\nabla^2 \boldsymbol{v} = -2\frac{v_0}{(d/2)^2}\,\hat{z}. \qquad (14.47)$$

For the geometry of the problem at hand, we may assume that

$$\boldsymbol{v} = v(t)\hat{\phi}. \qquad (14.48)$$

This is obtained by an average of **v** across the channel

$$\langle \eta \nabla^2 \boldsymbol{v} \rangle_{\text{channel}} = -R\rho\boldsymbol{v} \qquad (14.49)$$

satisfying div $\boldsymbol{v} = 0$, where R is a constant with dimensions $\sim v/d^2$.

Another simplifying assumption is that the temperature $T(\phi, t)$ in the fluid is taken uniform across the channel. The heat flow to and from the walls is given by

$$\langle \kappa \nabla^2 T \rangle_{\text{channel}} = -K[T(\phi, t) - T_w(\phi)] \qquad (14.50)$$

where K is constant with dimensions of inverse time. When the fluid is heated, its density decreases which provides a buoyant force so that it move up. Assuming a linear temperature dependence of density (Boussinesq approximation), we can write

$$\rho \mathbf{g} \approx \rho_0 (1 - \beta[T(\phi, t) - T_0])\,\mathbf{g} \qquad (14.51)$$

where β is the coefficient of thermal expansion and T_0 is some reference temperature. With these, the equations are

$$\rho_0 \left[\frac{\partial v}{\partial t} + (v.\nabla)v \right] = n - \nabla p + \rho_0[1 - \beta(T - T_0)]\mathbf{g} - R\rho_0 v,$$

$$\frac{\partial T}{\partial t} + v.\nabla T = -K[T(\phi, t) - T_w(\phi)]. \tag{14.52}$$

Let us now look for solutions:

$$v = v(t)\hat{\phi}, \quad p = p(\phi, t), \quad T = T(\phi, t). \tag{14.53}$$

Let dA be the area element of transverse section of the channel. Using

$$\hat{\phi}.\nabla = \frac{1}{\ell} \frac{\partial}{\partial \phi}, \quad \hat{\phi}.\frac{\partial \hat{\phi}}{\partial \phi} = 0. \tag{14.54}$$

The Navier-Stokes equation in (14.52) is projected on the area element $d\mathbf{A}$; different terms can be seen as

$$\int_A d\mathbf{A}.v = Av(t) := Q(t), \quad \text{flow rate,}$$

$$\int_A d\mathbf{A}.(v.\nabla)v = \int_A d\mathbf{A}. \left(v(t)\frac{1}{\ell} \frac{\partial(v(t)\hat{\phi})}{\partial \phi} \right) = 0,$$

$$\int_0^{2\pi} d\phi \frac{\partial p(\phi, t)}{\partial \phi} = 0. \tag{14.55}$$

The gravity terms is $\mathbf{g}.d\mathbf{A} = -gdA \sin \phi$ which integrates to zero when integrated from 0 to 2π.

The second term can be written as

$$(v.\nabla)T(\phi, t) = \frac{Q(t)}{A\ell} \frac{\partial T}{\partial \phi}. \tag{14.56}$$

Finally, the equations are

$$\frac{dQ}{dt} = \frac{\beta g A}{2\pi} \int_0^{2\pi} T(\phi, t) \sin \phi d\phi - RQ(t),$$

$$\frac{\partial T}{\partial t} + \frac{Q(t)}{A\ell} \frac{\partial T}{\partial \phi} = K[T_w(\phi) - T(\phi, t)]. \tag{14.57}$$

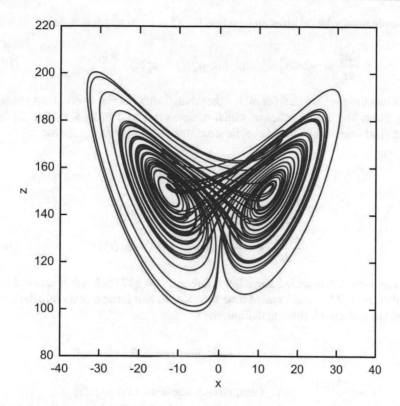

Fig. 14.3 Lorenz attractor obtained by solving (14.64) for parameter set: P = 10, r = 150, b = 1

For the wall temperature of the form, $T_w(\phi) := T_0 + T_1 \cos\phi$, $T_1 > 0$. Temperature, being periodic in ϕ, can be expressed in Fourier series:

$$T(\phi, t) = T_0 + a_0(t) + \sum_{n=1}^{\infty} [a_n(t) \cos(n\phi) + b_n(t) \sin(n\phi)].\qquad (14.58)$$

Substitute this in the temperature equation in (14.57):

$$\frac{da_0}{dt} = -Ka_0, \quad \text{implying } a_0(t) = a_0(0)e^{-Kt},\qquad (14.59)$$

which decays. For $n > 1$, we have

$$\frac{db_n}{dt} - \frac{nQ}{A\ell}a_n = -Kb_n, \quad \frac{da_n}{dt} + \frac{nQ}{A\ell}b_n = -Ka_n.\qquad (14.60)$$

A simple manipulation gives an equation for $s_n^2 = a_n^2 + b_n^2$ for $n > 1$:

$$\frac{da_n^2}{dt} = -2Ka_n^2, \quad \text{implying } a_n^2(t) = a_n^2(0)e^{-2Kt}. \tag{14.61}$$

Thus, we have shown that for all n other than 1, the Fourier coefficients vanish at long times. The only coefficient which survives is $n = 1$ in $T(\phi, t)$. For $Kt \gg 1$, (14.57) take the form which have the structure of the Lorenz equations:

$$\frac{dQ(t)}{dt} = \frac{\beta g A}{2}b_1(t) - RQ(t),$$

$$\frac{db_1(t)}{dt} - \frac{Q(t)}{A\ell}a_1(t) = K[-b_1(t)],$$

$$\frac{da_1(t)}{dt} + \frac{Q(t)}{A\ell}b_1(t) = K[T_1 - a_1(t)]. \tag{14.62}$$

We can now define scaled Rayleigh number, $r = g\beta T_1/2KR\ell$; a scaled Prandtl number, $P = R/K$, and a scaled time $\tau = Kt$. To cast these exactly into the Lorenz form, we make the following definitions:

$$x := \frac{Q(t)}{KA\ell} \qquad \text{(flow rate in the channel)}$$

$$y := \frac{rb_1(t)}{T_1} \qquad \text{(temperature variation at } \phi = \pi/2\text{)}$$

$$z := \frac{r(T_1 - a_1(t))}{T_1} \qquad \text{(temperature variation at } \phi = 0\text{).} \tag{14.63}$$

With these and some manipulations lead to the Lorenz equations [69]:

$$\dot{x} = -Px + Py,$$

$$\dot{y} = -y + rx - xz,$$

$$\dot{z} = -bz + xy, \tag{14.64}$$

where an overdot represents a time-derivative. The parameter, b is 1 for the present discussion (Fig. 14.3).

14.3   Nonlinear Wave Interaction

Consider nonlinear interaction among waves. If the amplitude of oscillations is not large and we use perturbation theory, to a leading approximation we arrive at a linear theory. Due to the principle of superposition, any perturbed state from the state of equilibrium can be expressed in terms of eigenmodes. Due to linearity of the description, all quantities are proportional to each other. Thus we may employ a scalar function, $\psi(\mathbf{r}, t)$ to describe any of these, and express it as

$$\psi(\mathbf{r}, t) = \sum_{\mathbf{k}} a_{\mathbf{k}} e^{-i\omega_{\mathbf{k}}t + i\mathbf{k}.\mathbf{r} + c.c.}. \tag{14.65}$$

For an elementary plane wave, $\psi_{\mathbf{k}} = a_{\mathbf{k}} e^{-i\omega_{\mathbf{k}}t + i\mathbf{k}.\mathbf{r}}$, we have

$$(\omega - \omega_{\mathbf{k}})a_{\mathbf{k}} = 0 \tag{14.66}$$

assuming that the dispersion relation $\omega = \omega_{\mathbf{k}}$ admits a solution for a non-zero $a_{\mathbf{k}}$.

If the wave amplitudes are not small, then we need to consider nonlinear terms into account. It is natural to consider first the quadratic terms proportional to $a_{\mathbf{k}'}a_{\mathbf{k}''}$ and $\sum_{\mathbf{k}'\mathbf{k}''}$ with condition equivalent to a conservation law, $\mathbf{k}' + \mathbf{k}'' = \mathbf{k}$. One may treat the left hand side of (14.66) as an operator $i\partial/\partial t$. In the case of an elementary plane wave, $\omega - \omega_{\mathbf{k}}$ transforms as

$$(\omega - \omega_{\mathbf{k}})a_{\mathbf{k}} e^{-i\omega_{\mathbf{k}}t + i\mathbf{k}.\mathbf{r}} = i\frac{\partial a_{\mathbf{k}}}{\partial t} e^{-i\omega_{\mathbf{k}}t + \mathbf{k}.\mathbf{r}}. \tag{14.67}$$

If quadratic terms are taken into account, (14.66) assumes the form

$$\frac{\partial a_{\mathbf{k}}}{\partial t} = \sum_{\mathbf{k}'} V_{\mathbf{k},\mathbf{k}',\mathbf{k}''} a_{\mathbf{k}'} a_{\mathbf{k}''} e^{-i(\omega_{\mathbf{k}'} + \omega_{\mathbf{k}''} - \omega_{\mathbf{k}})t} \tag{14.68}$$

where $\mathbf{k}'' = \mathbf{k} - \mathbf{k}'$. The real function, $V_{\mathbf{k},\mathbf{k}',\mathbf{k}''}$ is a "matrix element" of the wave interaction, uniquely determined by the nonlinear terms in the equation of motion. The quadratic terms,

$$a_{\mathbf{k}'} a_{\mathbf{k}''} e^{-i(\omega_{\mathbf{k}'} + \omega_{\mathbf{k}''})t + i(\mathbf{k}' + \mathbf{k}'').\mathbf{r}} \tag{14.69}$$

play the role of a power source with frequency, $\omega_{\mathbf{k}'} + \omega_{\mathbf{k}''}$ and wave vector $(\mathbf{k}' + \mathbf{k}'')$. This is accompanied by a similar power source with frequency, $\omega_{\mathbf{k}'} - \omega_{\mathbf{k}''}$ and wave vector $(\mathbf{k}' - \mathbf{k}'')$.

Three-wave processes provide an opportunity for eigenwave transformation and, therefore, lead to a complicated picture of energy transfer in the wave number phase space. These are called *triads*.

Let us consider the simplest case when there are only three waves with resonance condition,

$$\mathbf{k} = \mathbf{k}' + \mathbf{k}'', \qquad \omega_{\mathbf{k}} = \omega_{\mathbf{k}'} + \omega_{\mathbf{k}''}. \tag{14.70}$$

We denote the amplitudes as $a_1 = a_{\mathbf{k}}$, $a_2 = a_{\mathbf{k}'}$, $a_2 = a_{\mathbf{k}''}$ and corresponding frequencies as $\omega_1, \omega_2, \omega_3$. Assume that $\omega_1 = \omega_2 + \omega_3 > \omega_2 > \omega_3$. The equations for amplitudes (14.68) become

$$\frac{\partial a_1}{\partial t} = V_1 a_2 a_3 \tag{14.71}$$

for a_1 with $V_1 = V_{\mathbf{k},\mathbf{k}',\mathbf{k}''}$. Interchanging $\mathbf{k}$ and $\mathbf{k}'$ in (14.68), we have

$$\frac{\partial a_2}{\partial t} = V_2 a_1 a_3^*, \tag{14.72}$$

where $V_2 = V_{\mathbf{k}',\mathbf{k},-\mathbf{k}''}$. Further, substituting $\mathbf{k}''$ for $\mathbf{k}$ and $\mathbf{k}$ for $\mathbf{k}'$, we obtain

$$\frac{\partial a_3}{\partial t} = V_3 a_1 a_2^*, \tag{14.73}$$

with $V_3 = V_{\mathbf{k}'',\mathbf{k},-\mathbf{k}'}$. If we choose to normalize the amplitudes so that the energy $\mathcal{E}_{\mathbf{k}}$ of the kth wave is $\omega_{\mathbf{k}}|a_{\mathbf{k}}|^2$, then its momentum will be $\mathcal{P}_{\mathbf{k}} = \mathbf{k}|a_{\mathbf{k}}|^2$ and the quantity $\mathcal{N}_{\mathbf{k}} = |a_{\mathbf{k}}|^2$ can be interpreted as the number of waves.

Due to the conservation laws of energy and momentum, the above equations and their conjugates can be simply manipulated to give

$$\omega_1 V_1 + \omega_2 V_2 + \omega_3 V_3 = 0,$$

$$\omega_1 N_1 + \omega_2 N_2 + \omega_3 N_3 = \text{constant}, \quad \text{(energy)}$$

$$\mathbf{k}_1 N_1 + \mathbf{k}_2 N_2 + \mathbf{k}_3 N_3 = \text{constant}, \quad \text{(momentum)}. \tag{14.74}$$

The matrix elements obey following symmetry requirements:

$$V_{\mathbf{k},\mathbf{k}',\mathbf{k}''} = V_{\mathbf{k},\mathbf{k}'',\mathbf{k}'} = V_{-\mathbf{k},-\mathbf{k}',-\mathbf{k}''} = -V_{\mathbf{k}',\mathbf{k},-\mathbf{k}''} = -V_{\mathbf{k}'',\mathbf{k},-\mathbf{k}'}. \tag{14.75}$$

The equation for amplitudes of the interacting waves take the form

$$\frac{\partial a_1}{\partial t} = V a_2 a_3, \quad \frac{\partial a_2}{\partial t} = -V a_2 a_3^*, \quad \frac{\partial a_3}{\partial t} = -V a_1 a_2^*. \tag{14.76}$$

From these, it follows that

$$\frac{\partial}{\partial t}(|a_1|^2 + |a_2|^2) = 0 \quad \text{which implies that} \quad N_1 + N_2 = \text{constant}. \tag{14.77}$$

Similarly, $N_1 + N_3$ is constant. These are called the Manley-Rowe conservation laws.

14.4 Weak Wave Turbulence

A weakly nonlinear system of waves of small amplitudes $a_\mathbf{k}$ can be described in terms of a Hamiltonian where the wave amplitudes are employed as normal canonical variables [132]. For example, for energy density, $\rho v^2/2 + E(\rho)$, with

$$\mathbf{v_k} = \sqrt{\frac{ck}{2\rho_0}}(a_k - a_{-k}^*)\mathbf{k}, \qquad \rho_\mathbf{k} = \sqrt{\frac{\rho_0}{2ck}}(a_k + a_{-k}^*)k, \qquad (14.78)$$

we may expand energy density as a sum of terms with interaction matrix elements appear as smaller contributions to the quadratic non-interacting term. In general, the Hamiltonian in a general form is

$$H = \int \omega_k |a_\mathbf{k}|^2 + \int (V_{123}\, a_1 a_2^* a_3^* + c.c.)\delta(\mathbf{k}_1 - \mathbf{k}_2 - \mathbf{k}_3)d\mathbf{k}_1 d\mathbf{k}_2 d\mathbf{k}_3 + O(a^4). \qquad (14.79)$$

The condition of weak turbulence is: $\xi_k := |V_{\mathbf{kkk}} a_\mathbf{k}|k^3/\omega_k \ll 1$. The equation of motion of $a_\mathbf{k}$ is

$$\frac{\partial a_\mathbf{k}}{\partial t} = -i\frac{\partial H}{\partial a_\mathbf{k}^*} + f_k(t) - \gamma_k a_\mathbf{k} \qquad (14.80)$$

where f_k and γ_k denote forcing and damping. We assume that the forcing function is Gaussian random such that the spatial average, denoted by angular brackets, of the correlation function reads as:

$$\langle f_k(t) f_{k'}^*(t') \rangle = F(k)\delta(\mathbf{k} + \mathbf{k}')\delta(t - t') \qquad (14.81)$$

where for simplicity, we assume that $F(k)$ is non-zero only around some k_f. If we assume that the statistics of the wave system is close to Gaussian, the two-point correlation function rules the hierarchy of all correlations. The spatial structure is described by the correlation function: $\langle a_\mathbf{k}(t) a_{\mathbf{k}'}^{\prime*}(t') \rangle = n_k(t)\delta(\mathbf{k} + \mathbf{k}')$. Because dynamic equation (14.80) contains a quadratic nonlinearity, the equation of second moment contains third moment, third contains fourth, and so on, landing us with what is know as the "closure problem". In the approximation of weak turbulence in the inertial range, statistics is close to Gaussian and hence, fourth moment can be written as a product of two second moments. We get a closed equation [132]:

$$\frac{\partial n_k}{\partial t} = F_k - \gamma_k n_k + I_k^{(3)} \qquad (14.82)$$

where

$$I_k^{(3)} = \int (U_{k12} - U_{1k2} - U_{2k1}) d\mathbf{k}_1 d\mathbf{k}_2,$$

$$U_{123} = \pi [n_2 n_3 - n_1(n_2 + n_3)] |V_{123}|^2 \delta(\mathbf{k}_1 - \mathbf{k}_2 - \mathbf{k}_3) \delta(\omega_1 - \omega_2 - \omega_3).$$
$$(14.83)$$

Equation (14.82) is called kinetic equation for waves.

Let us define dissipation wave number k_d where the inverse time of nonlinear interaction,

$$|V_{\mathbf{kkk}}|^2 \frac{n_k k^d}{\omega(k)} \approx \gamma(k_d). \qquad (14.84)$$

We assume the nonlinearity to dominate over dissipation at k_d. Wave turbulence occurs when $k_d \gg k_f$ which is analogous to $Re \gg 1$.

In $I_k^{(3)}$, $\delta(\omega_1 - \omega_2 - \omega_3)$ implies that we account for only resonant processes which conserve the quadratic part of energy, $E = \int \omega_k n_k d\mathbf{k} = \int E_k d\mathbf{k}$. It can be shown that for the cascade picture to hold, the collision integral has to converge the inertial interval. Multiplying (14.82) by ω_k and integrating over inertial interval of $k_f \ll k \ll k_d$, the energy flux across a spherical surface is constant:

$$P_k = \int_0^k k^{d-1} dk \int d\Omega \omega_k I_k^{(3)} = \int \omega_k F_k dk = \int \gamma_k E_k dk = \epsilon. \qquad (14.8)$$

Equation (14.82) determines n_k.

Assume that the medium is isotropic and that we concentrate on the regime where scaling holds. The medium is then scale invariant:

$$\omega_k = k^\alpha, \quad \text{and}$$

$$|V_{\mathbf{k}_1 \mathbf{k}_2 \mathbf{k}_3}|^2 = V_0^2 k^{2m} \chi(\mathbf{k}_1/k, \mathbf{k}_2/k), \qquad (14$$

$\chi \approx 1$.

14.5 Turbulent Heat Flow: Structures and Scaling

We have discussed Rayleigh-Bénard convection in Chap. 11. With the increased heating rate, the flow becomes turbulent. To understand the nature of convective turbulence, geometrical structures and scaling ideas are found to be very useful. isolate features of turbulent flows, one adopts following strategies [70]:

1. Look for the qualitative geometry of typical recurrent structures,
2. Analyze fluctuations and obtain probability distributions,

3. Obtain qualitative characterizations of average flows,
4. Measure and study space-time dependence of velocity and other observables.

As summarized by Kadanoff [70], the most important lesson is that a complex system in a nonequilibrium state self organizes. Quoting Kadanoff, "...there are no separate laws of complexity science ... we have learned that theory of everything cannot include every interesting thing." An accessible summary may be found in [71].

In recent times, numerical computations of Navier-Stokes equation for a stirred fluid in a periodic box have shown the importance of new ideas like multiscale interactions, see [44].

Some recent ideas that have emerged from analysis of inverse energy cascades are scale invariance [18] and conformal invariance [13]. Under conformal transformations, the angles between vectors remain the same while their lengths are re-scaled. Conformal invariance has been found by analyzing the large-scale statistics of the boundaries of vorticity clusters [18], i.e. large-scale zero vorticity lines which are just the nodal curves. In a certain limit, these curves belong to the Schramm-Loewner Evolution or SLE curves. For an accessible description where the nodal curves are also presented in the context of quantum billiards, see [43].

Exercise 14.1 Estimate the Reynolds number for a cricket ball which is bowled at about 144 km/h, and, a table tennis ball which is driven by a fast top spin return at 45 km/h. Masses and diameters of the cricket and table tennis balls are $m_C \sim 160$ g, $d_C \sim 7$ cm, $m_{tt} \sim 2.5$ g, $d_{tt} \sim 3.8$ cm. From the Reynolds number, find the drag coefficient.

Exercise 14.2 Find the velocity flow between two planes, inclined to each other by an angle α. Discuss the case when the Reynolds number is high.

Superfluid Hydrodynamics and Quantum Turbulence

15

> *When we'd put rats in a maze, there was always one whose*
> *behaviour would contradict the theory, who couldn't have cared*
> *less about our clever hypotheses.*
>
> —Olga Tokarczuk

The term "Quantum turbulence" was introduced by R. J. Donnelly in a symposium dedicated to the memory of G. I. Taylor [25]. Quantum fluids like superfluids, superconductors, Higgs fields etc. may exhibit turbulent states [104] distinct from classical fluids due to the presence of off-diagonal long-range order [4, 129], measured in terms of an order parameter. Vorticity in superfluids exhibits topological defects in the order parameter which are line-like structures, the quantized vortices. The quantization of vorticity around a topological defect is a result of continuity in the order parameter. Feynman [32] argued that turbulence in ^{4}He exhibits a tangle of interacting quantized vortices, in contrast to a continuous distribution of vorticity classically.

We have discussed classical turbulence in the last chapter. The two main results we would like to keep in mind as we enter into this new subject are as follows:

1. Spectral density, $E(k) = c_K \epsilon^{2/3} k^{-5/3}$ where c_K is the Kolmogorov constant and ϵ is the average energy dissipation rate per unit mass.
2. Kolmogorov length scale, $\eta = (\nu^3/c)^{1/4}$.

15.1 The Superfluidity of Helium II

Kapitsa discovered the phenomenon of superfluidity in liquid helium in 1938. As a gas, it is inert, atoms interacting very weakly, and due to this, it turns from gas to liquid at 4.2 K at atmospheric pressure without becoming a solid. It can be solidified at approximately 25 atmospheres at a temperature close to zero Kelvin as the interatomic distances then enhance the interaction. From 4.2 K to 2.2 K, liquid helium behaves like an ordinary, normal liquid (called helium I). At 2.2 K, it turns into a superfluid, helium II, the behaviour of which is counter-intuitive and its understanding demands a number of new ideas.

The first property to discuss for a fluid is its viscosity. There are at least two methods of determining the coefficient of viscosity for ordinary fluids.

1. Measuring the flow of liquid through a narrow capillary tube under gravity is one way. The flow velocity is different points of a cross-section of capillary due to internal friction, it being maximum in the middle and decreasing toward the walls. We can determine the coefficient of viscosity by measuring the amount of liquid passing through capillary per unit time.
2. We can study the damping of torsional oscillations of a disc immersed in the liquid. Close to disc, fluid experiences drag from the disc which reduces with distance. Different layers of fluid, moving at different velocities, exert an internal friction on each other and generate heat. The dissipated energy damps the torsional oscillations, and the damping times is related to the coefficient of viscosity.

Usually, heat transfer occurs via heat flow or convection. *Heat flows* from one point to another in response to a temperature gradient. Rate of flow is determined by thermal conductivity which is the ratio of the heat to the temperature difference. A larger value of thermal conductivity implies that the heat will flow for small temperature difference.

Convection occurs when heat is transferred by actual motion of the fluid. Convection is associated with larger heat transfer. In He II, slightest temperature difference produces large heat transfer. We may neglect convection due to smallness of temperature difference, but then it implies an infinite thermal conductivity. Or, we conclude that heat transfer in He II is always caused by convection and that the temperature differences are absent. What is true? Now if we place a movable object appropriately, motion of the liquid can be verified accompanying heat transfer. It is experimentally verified that convection occurs indeed.

When any substance is cooled, the number of degrees of freedom are diminished. As gas or liquid becomes a solid on lowering the temperature, the ionic motion reduces to oscillations about their equilibrium positions. This intuitive thought gives way to the idea of elementary excitations, proposed by Landau. These are not motion of individual Helium atoms but actually collective motions of the atoms. These are sound vibrations. Collectivity arises due to the effect of oscillation of one atom on the other. The oscillations thus communicate across the substance. The whole solid or liquid vibrates, these constitute sound. Liquid helium is special—it implies that thermal motion is due to the presence of sound waves that can propagate in all directions. Thus, energy of the sound waves increases with temperature as the internal energy also increases.

Usual sound wave propagation is accompanied by a transfer of mass in the same direction. Sound wave can carry momentum which is exchanged when they hit a wall, for instance, resulting in what is termed as "sound pressure". Mass transfer occurs in case the sound waves, under some conditions, propagate in a specific direction rather than in all random directions. We have discussed that a body may generate sound when it is moving at a supersonic speed even though it does not vibrate. Thus if the body is a rest and a liquid moves at a sub-sonic speed, there is no emission of sound.

To understand He II's superfluidity, we will see that the observed phenomena results from the properties of sound waves in liquids. As a liquid flows through a capillary, the particles in the liquid interact with rough corrugation on the walls, causing the average kinetic energy of the liquid to increase and express as heat. This energy transfer would cause the emission of sound if the flow is supersonic. From the frame of reference moving with the liquid, the wall is moving in the opposite direction. Superfluidity means that the walls do not exert a drag on the fluid. They interact with the sound waves. There is an exchange of momentum between them and a privileged direction of sound propagation leading to a mass transfer occurs in the liquid. Drag on the liquid by the walls causes the mass transfer even though this mass of the liquid involved in the flow is small because the energy of sound waves is very small at low temperatures.

He II is visualized as possessing two independently moving components. The motion of one of them is superfluid—not accompanied by friction and the other, normal which exerts a drag on the walls. It should be emphasized that all helium atoms participate in both motions. As temperature increases, energy of the sound waves increases, and the mass of normal component increases until the superfluid component disappears entirely.

There is no pressure exerted by the superfluid component on a body moving in the He II at sub-sonic speed. If there is resistance, then we need to work on the body to move it—this work will turn into heat and lead to sound emission at sub-sonic speed—a contradiction. However, the normal component of the fluid exerted pressure.

We return to the methods discussed in the beginning.

1. *First method:* We do not find any viscosity in He II because superfluid component flows out rapidly. Normal fluid also flows out of the capillary, albeit slowly. So, in this method the measurement of the viscosity of the superfluid component is measured.
2. *Second method:* The rotating disk gives a non-zero value of viscosity as the disk moves in a fluid that has two components. The oscillations are damped due to the normal component.

Now we see why heat transfer in He II is accompanied by convective motion of the normal component. In the case of heat transfer in a vessel filled with He II, there is no mass transfer in one privileged direction because the pressure difference due to heat transfer generates flow of the superfluid component in the opposite direction. The velocity of superfluid motion is such that there is no mass transfer and the level of liquid remain the same.

When liquid helium in contact with its vapour is cooled below 2.17 K, it exhibits new properties in a new phase, He II. The specific heat about this temperature shows a dependence like $\log |T - T_\lambda|$.

The paradox mentioned above was explained by Tisza [116] by introducing the two-fluid model where liquid helium was thought to be a mixture of two interpenetrating fluids—one being viscous normal and the other being a superfluid. Landau [53] gave it a hydrodynamic description with velocities $(\mathbf{v}_n, \mathbf{v}_s)$. The superfluid accounts for rapid flow where $\mathbf{v}_s$ is irrotational. The phenomena related to liquid helium falls in two categories—(1) the irrotational $\mathbf{v}_s$ mentioned just now, and (2) when the vorticity of the superfluid is non-zero.

Helium remains in liquid state because the inter-atomic forces are weak and the zero-point motion is large, since the atomic mass is small, to keep it fluid even at absolute zero [67, 68]. Feynman [30] expounded that a lot of phenomena is due to the scarcity of available low-energy excited states in the Bose liquid. Although there exist excited states like phonons, low quantum of energies can be excited only by low wavenumber excitations, which correspond to long wavelengths. But the wavefunction cannot depend on the large-scale changes in the configuration. The large-scale motion consistent with no change in density is permutation of positions occupied by the atoms, consistent with the Bose-Einstein statistics. Landau developed his theory on this basis. The normal fluid consists of, in fact, quasi-particles like phonons and rotons, experimentally amenable by inelastic neutron scattering. Thermodynamically, the entropy of He II is entirely carried by the normal component. Pressure and temperature gradients are related to entropy by $\Delta P / \Delta T = \rho S$.

To understand another novel excitation, let us consider a vessel containing normal component and another component containing superfluid component, connected by a channel. The part with normal component is connected to a heat reservoir, and is at temperature $T + \Delta T$ whereas the superfluid part is at a temperature, T. The channel is seen to draw superfluid component to the heater, and since there is no net mass flux, normal fluid must counterflow. If the heater is turned on and off periodically, the motion of the two fluids sets up a longitudinal standing wave of *second sound*. While the first or usual sound is due to density fluctuations, the second sound is due the two-fluid model consisting of temperature or entropy fluctuations.

15.2 Quantum Fluid Mechanics

Theoretical description with two fluids can be given for small velocities by the equations of motion:

$$\rho_s \frac{D\boldsymbol{v}_s}{Dt} = -\frac{\rho_s}{\rho}\nabla p,$$

$$\rho_n \frac{D\boldsymbol{v}_n}{Dt} = -\frac{\rho_n}{\rho}\nabla p - \rho_s S\nabla T + \eta\nabla^2\boldsymbol{v}_n,$$

$$\nabla \times \boldsymbol{v}_s = 0. \tag{15.1}$$

The total density $\rho = \rho_s + \rho_n$. D/Dt denotes the convective derivative. The terms $\rho_s S\nabla T$ represent the fountain pressure (in case of the famous experiment). When $\boldsymbol{v}_n = \boldsymbol{v}_s$, the two equations add up to a Navier-Stokes equation for the total fluid. Ordinarily $\boldsymbol{v}_n \neq \boldsymbol{v}_s$ because of the irrotational condition put forward by Landau. However, experiments showed that (15.1) is incomplete, and that a description of quantized vortices needs to be incorporated.

15.2.1 Quantized Vortices

The quantization of vortices was first taught to a class in 1946, and announced in a typical terse manner by one of the most influential physicists, Lars Onsager in a conference in Florence in 1949. The superfluid component is described by an order parameter which takes the form of a complex field,

$$\psi = a\, e^{i\theta} \tag{15.2}$$

with a, θ as real fields. The long-range order in quantum fluids is captured by the order parameter which indicates the level of synchronization of the atomic wavefunctions at a long range. Superfluid density is

$$\rho_s = m\,\psi^*\psi = ma^2 \tag{15.3}$$

where m is the mass of fluid atom. As $T \to 0$, ρ_s/ρ increases. Quantum mechanical mass current,

$$\mathbf{j}_s = -\frac{\hbar}{2}\left(\psi^*\nabla\psi + (\nabla\psi^*)\psi\right) = ha^2\nabla\theta. \tag{15.4}$$

Superfluid velocity,

$$\boldsymbol{v}_s = \frac{\mathbf{j}_s}{\rho_s} = \frac{\hbar}{m}\nabla\theta \tag{15.5}$$

which is well-defined if $a \neq 0$. Obviously, since $\nabla \times \boldsymbol{v}_s = 0$, superfluid flow is a potential flow in a simply-connected region of the fluid. Multiply-connected regions are caused by the presence of line-like defects which either form loops or end on the boundaries. The gradients of phase, $\nabla \theta$ causes these defects. Onsager noted that continuity of order parameter ψ requires that the phase change along a closed contour will be

$$\oint_C \nabla \theta . d\boldsymbol{\ell} = 2\pi n, \quad n \in Z_+. \tag{15.6}$$

Circulation,

$$\Gamma = \oint_C \boldsymbol{v} . d\boldsymbol{\ell} = n . \frac{h}{m} := n\kappa \tag{15.7}$$

where κ is the quantum of circulation. The value of h/m is 9.97×10^{-4} cm^2/s in He II by taking m as mass of a bare helium atom. This is surprising if we note that the atoms in a liquid are strongly coupled.

Feynman [31] suggested that vorticity is constrained to the cores of quantized vortices, thin as a single fluid atom of mass m. With radius of the core denoted by 'a' and the mean distance between the vortices denoted by 'b'. To get a reasonable estimate of the energy contained in a vortex line, let us consider an isolated line along the axis of a cylinder of length L, radius b. The velocity at radius r is $\hbar/mr$, the kinetic energy is

$$\frac{1}{2} \int \rho_s m (\hbar/mr)^2 2\pi r dr . L. \tag{15.8}$$

The upper limit of the integral is b and the lower limit is the radius of the core, a. Therefor the energy needed to form a line, per unit length, is

$$E_v = \rho_s \pi \frac{\hbar^2}{m} \hbar^2 \ln \frac{b}{a} \approx \frac{\rho_s \kappa^2}{4\pi} \ln \frac{b}{a}. \tag{15.9}$$

This is a lot of energy: with $b/a \sim 10^7$, it is about 1.85×10^{-7} erg/cm. The centrifugal force on each ring of fluid surrounding the core is balanced by the radial pressure gradient,

$$\frac{dp}{dr} = \frac{\rho_s v_s^2}{r} = \frac{\rho_s \kappa^2}{4\pi^2 r^3}. \tag{15.10}$$

Feynman observed that since energy is proportional to κ^2, a double-quantized vortex line would have four times more energy, hence it would be relatively unstable. Moreover, he noted that the number of flux lines arranged parallel to the axis of rotation (for instance, of a rotating of Helium with an angular velocity Ω) would be roughly uniform, given by

$$n_0 = \frac{\nabla \times \boldsymbol{v}_s}{\kappa} = \frac{2\Omega}{\kappa} = 2000\,\Omega, \qquad (15.11)$$

a large number.

In general, the normal fluid is described by the Navier-Stokes equation and the superfluid by the Gross-Pitaevskii equation [89] at $T = 0\,K$. To each equation, coupling terms are added to describe the mutual friction between the two fluids, as initially observed and described in classic work from doctoral thesis by Vinen [120–122]. The equations without the irrotational condition (curl $\boldsymbol{v}_s = 0$) become

$$\rho_s \frac{D\boldsymbol{v}_s}{Dt} = -\frac{\rho_s}{\rho}\nabla p + \rho_s S\nabla T - \mathbf{F}_{ns},$$

$$\rho_n \frac{D\boldsymbol{v}_n}{Dt} = -\frac{\rho_n}{\rho}\nabla p - \rho_s S\nabla T + \mathbf{F}_{ns} + \eta\nabla^2\boldsymbol{v}_n. \qquad (15.12)$$

As mentioned earlier, second sound is attenuated by the presence of quantized vortices. The attenuation coefficient, α is obtained by assuming that the mutual friction can be written as

$$\mathbf{F}_{ns} = -B\frac{\rho_s + \rho_n\omega}{2\rho}(\boldsymbol{v}_n - \boldsymbol{v}_s) \qquad (15.13)$$

where vorticity, ω is κL. In turbulence experiments, line density L, 0–1,500,000 is observed. Through a rotating bucket carrying He II, a beam of ions produced by an α-emitting radioactive source is passed. Negative ions get stuck to quantized vortices at temperatures below 1.7 K. By applying an electric field, the missing ions are moved up and collected at the top. A further analysis gives the spatial information on the line density. Thus, vorticity can be measured.

15.2.2 Dynamics of Quantized Vortex Rings

Donnelly and Roberts [24] postulated that He II contains a distribution of quantized vortices of all shapes and sizes, oriented randomly but carried on the superfluid. Since the smallest of these rings would be of the size of 0.1 nm, the energy associated with these rings will be large due to quantum localization effects. Corresponding to minimum energy E_m of a ring, we may associate a momentum p_m. The number of such rings can be calculated by the methods of statistical mechanics, and, the population of states beyond E_m would be given by including thermal collisions. This calculation coincides with the calculation of density of rotons, as expected by Feynman [31]. The role of momentum of vortex rings is played by the impulse,

$$p := \rho_s \kappa \pi R^2, \quad \text{or} \quad R = \sqrt{p/\rho_s \kappa \pi}. \tag{15.14}$$

Energy of the vortex ring has been shown by Roberts and Donnelly [99],

$$E = \frac{1}{2} \rho_s \kappa^2 R[\log(8R/a) - 3/2]. \tag{15.15}$$

The velocity of a vortex ring of radius R is then given by Hamilton's equation:

$$v = \frac{\partial E}{\partial p} = \left(\frac{\kappa}{4\pi R} \right) \left[\log \left(\frac{8R}{a} \right) - \frac{1}{2} \right]. \tag{15.16}$$

Vortex ring calculation was generalized by Arms and Hama [3] where an approximate calculation of the motion of arbitrary configurations of very thin vortex lines is carried out. An analogy with Biot-Savart law is used for velocity at some point in space induced by a vortex line. Analogy of velocity (vorticity) with magnetic field $\mathbf{H}$ (current density, $\mathbf{j}$) may be noted: "$\boldsymbol{\omega} = \nabla \times \boldsymbol{v}$" is similar to "$\mu_0 \mathbf{j} = \nabla \times \mathbf{H}$". Let us consider a space curve representing a vortex line with position vector for a point on the line, $s(\xi, t)$ (Fig. 15.1). Let us denote the first and second derivatives of s w.r.t. ξ by s' and s'' respectively. The magnitude of s'' is $1/R$, the local curvature, and, s' is along $\hat{\kappa}$, unit vector along the vortex line. The vector, $s' \times s''$ is approximately in the direction of the local induced velocity v, with magnitude $1/R$. The instantaneous velocity of the line is

$$\boldsymbol{v}_L = \frac{d}{dt} s(\xi, t) = \dot{s}. \tag{15.17}$$

The local induced velocity,

$$v = \frac{\kappa}{4\pi} \int_{\text{line segments}} (s_0 - \mathbf{r}) \times \frac{ds_0}{|s_0 - \mathbf{r}|^3} \tag{15.18}$$

where $\mathbf{r}$ is any point in the fluid. Integral in (15.18) diverges if $\mathbf{r} = s_0$ is a point on the line. Expanding s in Taylor series about s_0, (15.18) becomes

$$v \approx \frac{\kappa}{4\pi} \int \frac{d\xi}{2\xi} s' \times s''. \tag{15.19}$$

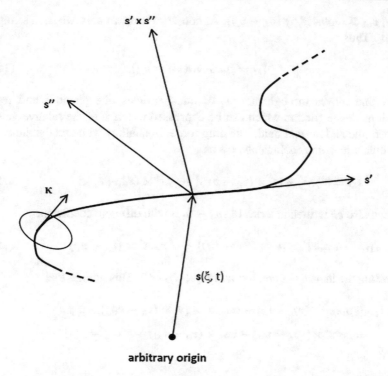

Fig. 15.1 The space curve represents a vortex line with position vector $s(\xi, t)$. The derivative, $ds/d\xi$ and curvature, $d^2s/d\xi^2$ are denoted by s' and s'' respectively; whereas κ denotes the unit vector along the vortex line. The cross product is approximately along the local induced velocity

Ignoring "nonlocal" portions, more than distance L away from s_0, approximating the cross product by its value at s_0, we get local self-induced velocity

$$v_i = \frac{\kappa}{4\pi} \log\left(\frac{L}{a}\right) s' \times s''$$

$$\approx \frac{\kappa}{4\pi R} \log\left(\frac{L}{a}\right). \tag{15.20}$$

This tells us how a vortex ring moves through a fluid. A vortex ring moves through the fluid because of curvature. If $a = 1$ Å, $R = 10^{-3}$ cm, velocity is about 1 cm/s.

Analogous to the lift force on a cylinder we studied in aerodynamics, a force on vortex lines is its generalization. With circulation κ and the velocity of the flow, v_∞, the force is $f_M = \rho v_\infty \times \kappa$. This expression from aerodynamics is valid for vortex line at absolute zero with the condition that it can move at a velocity v_L different from local superfluid velocity. Denoting the sum of flow v_s and self-induced flow v_i

by v_{sl}, v_∞ is replaced by $(v_L - v_{sl})$. At non-zero temperatures, we need to replace ρ by ρ_s. Thus,

$$\mathbf{f}_M = \rho_s \kappa s' \times (v_L - v_{sl}). \tag{15.21}$$

Vortex line moves through the gas of quasiparticles like phonons and rotons. Collisions create friction which can be expressed in terms of the relative velocity between line and normal fluid. The drag force, in analogy with aerodynamics, can be written with some coefficients, γ's as

$$\mathbf{f}_D = -\gamma_0 s' \times [s' \times (v_n - v_L)] + \gamma_0' s' \times (v_n - v_L). \tag{15.22}$$

This can also be written in terms of $(v_n - v_{sl})$ with different coefficients:

$$\mathbf{f}_D = -\alpha \rho_s \kappa s' \times [s' \times (v_n - v_{sl})] - \alpha' \rho_s \kappa s' \times (v_n - v_{sl}). \tag{15.23}$$

Neglecting the inertia of core, the sum $\mathbf{f}_M + \mathbf{f}_D = 0$. This implies that

$$\begin{aligned}
\mathbf{f}_M + \mathbf{f}_D &= \rho_s \kappa s' \times (v_L - v_{sl}) - \alpha \rho_s \kappa s' \times [s' \times (v_n - v_{sl})] - \alpha' \rho_s \kappa s' \times (v_n - v_{sl}) \\
&= \rho_s \kappa s' \times [(v_L - v_{sl}) - \alpha s' \times (v_n - v_{sl}) - \alpha'(v_n - v_{sl})] \\
&= 0. \tag{15.24}
\end{aligned}$$

Thus the term in the square bracket must be in direction of s'. or, it must be zero. Considering the truth of the latter, we have

$$v_L = v_{sl} + \alpha s' \times (v_n - v_{sl}) - \alpha' s' \times [s' \times (v_n - v_{sl})]. \tag{15.25}$$

This is a fundamental result in vortex dynamics.

15.2.3 Uniformly Rotating He II

The first application of mutual friction proposed by Vinen was in understanding attenuation of second sound in uniformly rotating He II. The frictional force is applied on an isolated vortex line. Rapid rotation of a container will produce a uniform array of quantized vortices. In equilibrium, sum of force on the normal fluid and drag force per unit volume must be zero:

$$\mathbf{F}_{ns} + \mathbf{f}_D n_0 = 0. \tag{15.26}$$

Using this and (15.23),

$$\mathbf{F}_{ns} = \omega \rho_s \alpha s' \times [s' \times (v_n - v_s)] \tag{15.27}$$

where the magnitude of vorticity, ω is 2Ω. Second-sound resonance experiments give values of α and α'.

A more general case of mutual friction arises when vortex lines are curved. The local induced velocity, given by the Arms-Hama equation, $v_i = (\kappa/4\pi R)\ln(L/a)$, is non-zero. Its direction will be along $s' \times s''$:

$$s' \times s'' = (s'.\nabla)s'. \tag{15.28}$$

We may write an approximate expression, $v_i \approx v(s'.\nabla)s'$, and using (15.9), we have

$$v = \frac{\kappa}{4\pi}\ln\frac{L}{a} = \frac{E_v}{\rho_s\kappa}. \tag{15.29}$$

If the curved lines are dense and form a continuum, then (15.25) and (15.26) can be generalized to

$$\mathbf{F}_{ns} = \rho_s\omega\alpha s' \times [s' \times (v_n - v_s)] + \rho_s\omega\alpha' s' \times (v_n - v_s)$$
$$+ \rho_s\omega\alpha v\nabla \times s' + \rho_s\omega\alpha' v(s'.\nabla)s'. \tag{15.30}$$

Last two terms arise due to v_i, and the identity $(s'.\nabla).s' = -s' \times (\nabla \times s')$ has been used. At absolute zero, $v_L = v_{sl}$ which we now call simply v_s. The superfluid equation of motion becomes

$$\frac{\partial v_s}{\partial t} = v_s \times \omega + v\omega(s'.\nabla)s' + \nabla\mu. \tag{15.31}$$

Together, $v_s \times \omega$ and $\partial v_s/\partial t$, becomes total derivative, we get

$$\rho_s\frac{dv_s}{dt} = -\frac{\rho_s}{\rho}\nabla p + \rho_s S\nabla T + v\omega(s'.\nabla)s' - \mathbf{F}_{ns}. \tag{15.32}$$

Normal and superfluid components of velocity follow equations, called Hall-Vinen-Bekarevich-Khalatnikov equations [81]. These equations have been used to understand stability of Taylor-Couette flow in He II [8,113]. For an array of vortices, there are other excitations [23]

15.3 Quantum Turbulence

Turbulence in superfluid is dominated by a complex, interacting tangle of quantized vortices. The relevant scales are as follows:

1. *Length scales*: smallest scale is the diameter of quantized vortex core ($\sim 10^{-10}$ m) for He II; upper bound is the system size or many times the inter-vortex spacing ~ 1 cm.

2. *Time scales*: slowest time-scales correspond to long-range vortex-vortex interactions ~ 1 s; fastest time-scale corresponds to wave motion along the quantized vortices with periods $< 10^{-9}$ s.

The vortex waves are transverse, circularly polarized displacements by the kinetic energy per unit length of a quantized vortex and restored by the vortex tension. Such Kelvin waves along a rectilinear vortex have an approximate dispersion relation [124]:

$$\omega = \frac{\kappa k^2}{4\pi} \left(\log \frac{1}{ka_0} + c \right), \tag{15.33}$$

a_0 is the vortex cut-off parameter and the constant, $c \approx 1$. Additional complications arise due to interaction between the normal and superfluid components mediated by the quantized vortices. Two fluids couple, so turbulence in one can trigger turbulence in the other.

15.3.1 Landau Two-Fluid Model for a One-Component Fluid

To the Landau two-fluid model, the interaction of vortices has been introduced [12]. We can write an equation for the time evolution of a classical complex function ψ in the form of a generalized nonlinear Schrödinger equation (gNLSE) by including a nonlocal two-body interaction potential $V(|\mathbf{x} - \mathbf{x}'|)$:

$$i\hbar \frac{\partial \psi}{\partial t} = E(i\nabla)\psi + \int d\mathbf{x}' \, |\psi(\mathbf{x}')|^2 \, V(|\mathbf{x} - \mathbf{x}'|) \, \psi + \mu \, |\psi|^2 \, \psi. \tag{15.34}$$

$E(k)$ gives intrinsic dispersion of elementary excitations, given by $\hbar^2 k^2 / 2m$ for quadratic dispersion where mass of the particle is m. The chemical potential μ comes from the equation of state from where pressure P gets related by $d\mu(n) = dP(n)/n$. Assuming that the small perturbations in number density about a uniform density, n_0 which is the number of particles N occupying the volume $\mathcal{V}$. Defining Fourier transform of $V(r)$ by

$$\tilde{V}(k) = \int d\mathbf{r} \, V(r) e^{-i\mathbf{k}.\mathbf{r}}, \tag{15.35}$$

Fourier transform of (15.34) gives the dispersion law:

$$\omega^2 = \frac{\hbar^2}{4m^2} k^4 + c_0^2 k^2 + \frac{n_0}{m} \tilde{V}(k) k^2 \tag{15.36}$$

with the speed of sound, $c_0 = ((1/m)\partial P/\partial n_0)^{1/2}$. Pressure and energy are related by $P = n_0^2 \partial (E/N)/\partial n_0$. The gNLSE is a phenomenological model of superfluid helium because in helium there exists a strong coupling of highly occupied modes of low occupation due to strong interactions. Berloff et al. [12] interpreted gNLSE as a mathematical model that brings about Landau two-fluid model and treats vortices self-consistently. As a result, the complex function ψ is not associated with the condensate. Madelung transformation on ψ in (15.34) entails conservation of mass and "Newton's law in Eulerian description":

$$\frac{\partial n}{\partial t} + \nabla.n\boldsymbol{v} = 0,$$

$$m\frac{d\boldsymbol{v}}{dt} + m(\boldsymbol{v}.\nabla)\boldsymbol{v} = -\nabla\mu_0(n), \tag{15.37}$$

where

$$\mu_0(n) = \mu(n) + \int d\mathbf{x}'\, n(\mathbf{x}')V(|\mathbf{x} - \mathbf{x}'|) - \frac{\hbar^2}{2m^2}\frac{\nabla^2\sqrt{n}}{\sqrt{n}}. \tag{15.38}$$

The extra term in Euler equation is due to the presence of quantum pressure which differentiates superflow from perfect Euler flow. This term is responsible for vortex reconnections [134] etc. In the context of modelling the finite-temperature superfluid ^{4}He, due to the fact that it is a strongly interacting system, small occupation numbers are coupled with high occupation numbers. Thus, (15.34) cannot be used. We may consider evolution of the field

$$\psi(\mathbf{r}, t) = \sum_{\mathbf{k}} a_{\mathbf{k}} e^{i\mathbf{k},\mathbf{x}} \tag{15.39}$$

where $a_{\mathbf{k}}$ define the occupation numbers $n_{\mathbf{k}}$ by $\langle a_{\mathbf{k}} * a_{\mathbf{k}'} \rangle = n_{\mathbf{k}}\delta_{\mathbf{k}}\delta_{\mathbf{k}'}$. In equilibrium, the superfluid component corresponds to $\mathbf{k} = 0$. The normal component follows the classical limit of Bose-Einstein distribution,

$$n_{0,k} = \frac{k_B T}{\hbar^2 k^2/2m - \mu}. \tag{15.40}$$

The long-wavelength part of the Fourier spectrum corresponds to the superfluid component of the tangle of vortices. From here, equilibrium temperature can be determined. However, the actual case of interacting vortices demands many refinements, and constitutes the subject of intensive research.

Thermal dissipation in quantum turbulence, comprising tangled vortices, is studied [49] by solving the coupled system involving the Gross-Pitaevski equation (GP) and the Bogoliubov-de Gennes equation (BdG). It is assumed that dissipation in the system of quantized vortices is caused by the interaction between condensate and its excitations. The excitations are described by the BdG equation. Dissipation is studied as a model for mutual friction in dilute BECs by calculating the dynamics of straight vortex, and thereby obtaining the friction coefficients. The Hamiltonian operator for the BEC as a function of field operator $\hat{\Psi}$:

$$\mathbf{H} = \int \hat{\Psi}^{\dagger} \left(-\nabla^2 - \mu + \frac{g}{2} |\hat{\Psi}|^2 \right) \hat{\Psi} \tag{15.41}$$

where $\hat{\Psi}$ is the boson field operator, μ is the chemical potential, and g is the coupling constant. The dynamics is given by $i \partial \hat{\Psi} / \partial t = \mathbf{H} \hat{\Psi}$. $\hat{\Psi}$ can be written as a mean-field ansatz, $\hat{\Psi} = \Phi + \hat{\chi} + \hat{\zeta}$, conserving the particle number. Φ is the macroscopic wavefunction, $f e^{i\phi}$ where f^2 is the condensate density and the superfluid velocity $\boldsymbol{v} = 2\nabla\phi$. χ and ζ correspond to the first-order and higher-order excitations respectively. For completeness, let us write the equations satisfied by Φ and $\hat{\chi}$. Neglecting the higher-order excitations obtains the Gross-Pitaevski equation:

$$i \frac{\partial \Phi}{\partial t} = [-\nabla^2 - \mu - g(|\Phi|^2 + 2\langle \hat{\chi}^{\dagger} \hat{\chi} \rangle)]\Phi + g\langle \hat{\chi}^2 \rangle \Phi^*. \tag{15.42}$$

The first-order correction follows the BdG equation:

$$i \frac{\partial \hat{\chi}}{\partial t} = [-\nabla^2 - \mu + 2g|\Phi|^2]\hat{\chi} + g\Phi^2 \hat{\chi}^{\dagger}. \tag{15.43}$$

Written in components, a system of coupled equations are numerically solved [49], leading to remarkable configurations of quantized vortices. Of course, the tangle of vortices was predicted by Feynman, and developed by the work of Schwarz [101]. Numerical results from [49] represents the culmination of Feynman's intuition.

15.3.2 Kinetic Theory

In his treatment of superfluidity, Beliaev [11] separated the field operator as a sum of an ensemble-averaged condensate wavefunction and a residual fluctuating part:

$$\hat{\Psi}(\mathbf{r}, t) = \psi_c(\mathbf{r}, t) + \hat{\psi}'(\mathbf{r}, t) \tag{15.44}$$

where $\psi_c(\mathbf{r}, t) = \langle \hat{\Psi}(\mathbf{r}, t) \rangle$. A two-component wavefunction was written with $k = 0$ mode and excited modes. For this, kinetic theory was constructed by Kirkpatrick and Dorfman [46–48]. They could account for the Bogoliubov excitation spectrum for quasiparticles, and developed the Green-Kubo formula to explain damping in terms of transfer of particles in and out of the condensate. For inhomogenous condensates, this approach has been generalized [37, 133]. The development of equations for the two components has been built along the lines of two-fluid model, starting with the Heisenberg equations of motion:

$$i\hbar \frac{\partial \psi_c}{\partial t} = \left\{ -\frac{\hbar^2 \nabla^2}{2m} + V_0[n_c(\mathbf{r}, t) + 2n'(\mathbf{r}, t)] \right\} \psi_c(\mathbf{r}, t) - i R(\mathbf{r}, t)\psi_c(\mathbf{r}, t),$$

$$\frac{\partial f}{\partial t} + \frac{\mathbf{p}}{m}.\nabla_{\mathbf{r}} f - (\nabla_{\mathbf{r}}).(\nabla_{\mathbf{p}} f) = C_{12}[f, \psi_c] + C_{22}[f] \tag{15.45}$$

where

$$n_c = |\psi_c|^2, \quad n' = \int \frac{d\mathbf{p}}{h^3} f(\mathbf{p}, \mathbf{r}, t), \tag{15.46}$$

and the dissipation term with iR allows the transfer of particles in and out of the condensate while preserving the conservation laws of momentum and energy for binary collisions. Self-consistency requires $U = 2V_0(n_c + n')$, and $f(\mathbf{p}, \mathbf{r}, t)$ is the Wigner function,

$$f(\mathbf{p}, \mathbf{r}, t) = \int d\mathbf{r}' e^{i\mathbf{p}.\mathbf{r}'/\hbar} \langle \hat{\psi}'^{\dagger}((\mathbf{r} + \mathbf{r}')/2, t)\hat{\psi}'((\mathbf{r} - \mathbf{r}')/2, t) \rangle. \tag{15.47}$$

The equation satisfied by f is quantum Boltzmann equation for thermal particles. Importantly, the collisional terms here contain evolution of the distribution function with stimulated and spontaneous terms.

15.4 Comparing Classical and Quantum Turbulence

Recalling the length-scales discussed above, the largest scale can accommodate many quantized vortices. We have a situation similar to "large quantum numbers" when we discuss quantum-classical correspondence [62]. In this limit, we expect features of classical turbulence to appear. Importantly, there is an experimental support towards this, as summarized in [85].

In line with our discussion on basic concepts of classical turbulence, we now describe key concepts here.

1. *Energy cascades*: Quantum turbulent energy cascades is still a subject of active research. The difference in the cascade process is due to the fact that vortex cores are topologically constrained and there are no direct viscous losses [123]. The cascade sequence is: (a) quantized vortices interact with normal fluid which tends to align them. This bundle of vortices sets up large-scale motion corresponding to small wave numbers. (b) Reconnection between individual vortices leads to transmission of energy to larger wave numbers. (c) In addition, reconnections give birth to polychromatic Kelvin waves on the vortex lines. (d) The nonlinear

interaction of Kelvin waves produces even larger wave numbers until they lose energy to phonon emission, radiating energy to the boundaries [124].

2. *Large-scale motions*: Unlike turbulence in classical fluids, vorticity in the superfluid component neither diffuses nor forms coherent structures. However, groups of quantized vortices can mimic large-scale circulation as in classical fluids. Let us consider fluid contained in an infinitely long, cylindrical vessel rotating at an angular frequency Ω about the vertical axis. If the fluid inside the vessel is classical, then it will eventually reach a state of a solid-body rotation where each fluid element rotates with a velocity increasing linearly with radius of the cylinder, s.

    Vorticity in a quantum fluid is quantized; thus, it is clear that many quantized vortices aligned parallel to the rotation axis are required to produce a coarse-grained velocity field similar to the classical case. Flow around each quantized vortex is $\mathbf{v} = \kappa/(2\pi s)\hat{\phi}$. In contrast to the solid rotation, this is decaying with s. Feynman showed that the energy is lowest for the state where the quantized vortices are arranged in a triangular lattice arrangement. This corresponds to the lowest-energy approximation to the solid-body rotation [31]. This lattice structure has been experimentally observed in superfluid He II [130].

3. *Small-scale motions*: In classical fluids, small-scale motions occur on the scale of Kolmogorov length, η. Below this scale, viscosity facilitates diffusion of momentum and thereby smoothens the flow. Define small scales in a quantum fluid on lengths smaller than the spacing between quantized vortices, ℓ. Vorticity cannot diffuse, and thus it cannot have a homogenizing effect at the scale of 10^{-8} m. Below the scale of ℓ, vorticity undergoes extreme quantum fluctuations because it is topologically constrained to thin filaments in motion. In this way, this is distinctly different from classical case. Reconnection—a result of local interaction between quantized vortices—can produce velocities as large as the speed of sound in the fluid. Thus, small-scale dynamics is very different in quantum turbulence.

15.5 A Very Brief Summary of Some Experimental Methods

Velocity distribution functions, structure functions and energy spectra are some of the most important quantities one would like to infer about from the experiments. This is accomplished by measuring pressure-head fluctuations by using Pitot tubes [74]. Quantum turbulence is generated and detected by a number of small mechanical oscillators like spheres, wires, nanowires, quartz tuning forks (for details, see [105]. Typical methods like particle imaging and particle-tracking velocimetry have been used to observe individual quantized vortices and reconnections [14, 15]. Attenuation of second sound is the well-known measurement tool [120] in ^{4}He above 1 K, it gives the vortex-line density L—the total length of the quantized vortex line in a unit volume [7].

To make measurement of decaying quantum turbulence, negative ions are injected and manipulated by electric field. Above 0.8 K, bare ions dominate whereas below 0.7 K, they are self-trapped in the quantum vortex cores within 1 μm. Short pulses of ions are sent across the helium cell, the reduction of amplitude of pulses at the collector on the opposite end is measured [29, 128]. This is modelled by incorporating their interaction with quantized vortices to reveal the vortex line density.

15.6   A Few Questions

Drawing inspiration from a relatively recent review on quantum turbulence in quantum gases [72], close with an (incomplete) list of challenges for future.

1. The onset of turbulence, its decay and scaling laws as the interacting environment changes is an important subject. Can one "control" superfluid turbulence ?
2. Visualizing and characterizing vortex tangle in liquid Helium systems, determining kinetic energy spectrum, extracting momentum distribution from two-dimensional integrated density profiles of three-dimensional vortex tangles consists a challenging problem.
3. Vortex reconnection seems to play an important role in superfluid turbulence. What might be the finite-size modification in Kolmogorov turbulence, which originates from vortex bundles.
4. What are experimental observables which quantify quantum turbulence?
5. Most of the studies are confined to bosonic fluids or mixtures. Quantum turbulence in fermionic systems and ultracold atomic systems is important to understand.

Solutions

Problems of Chapter 1

Solution 1.1 The number density in the atmosphere is given by

$$n = n_0 \exp -\frac{h}{H},\tag{1}$$

where $n_0 = 2.5 \times 10^{19}\,\text{cm}^{-3}$ and $H = 8.5 \times 10^5\,\text{cm}$.

For fluid approximation, we need scale-height (H) to be much larger than the mean free path (λ). In the extreme case, we set both length comparable i.e $\lambda = (n\sigma)^{-1} \sim H$. Thus, as one goes up in the atmosphere the density fall exponential and eventually becomes comparable with the scale-height beyond which fluid approximation will fail miserably. Substituting numbers we get height to be approximately 200 km.

Solution 1.2 At an instant t, the water jet we see, is a streakline. It is formed by water particles that left the tank at consecutive instants. If we neglect surface tension and the friction between water and air, the path would be a parabola. Each particle flows out of the orifice with a different horizontal speed as it depends on H. H decreases with time.

Let us consider a coordinate system with origin at the orifice where x is the horizontal direction and y is vertical. Thus the ground has $y = -h$. The time $t = 0$ is when water flows out. The horizontal speed of the first water particle leaving the tank is governed by the Torricelli's law:

$$v_0 = \sqrt{2gH_0/(1 - (a/S)^2)}\tag{2}$$

© The Author(s), under exclusive license to Springer Nature Switzerland AG 2022

S. R. Jain et al., *A Primer on Fluid Mechanics with Applications*,

https://doi.org/10.1007/978-3-031-20487-6

where we distinguish the initial height by $H = H_0$ for clarity. The path of the particle is given by simple equations:

$$x = v_0 t, \quad y = -gt^2/2 \tag{3}$$

or, just $x^2 = -2v_0^2 y/g$. It takes a duration of $T = \sqrt{2h/g}$ to reach the ground, thus forming the complete streakline.

However, at an instant T, the particle leaving the tank has a horizontal speed,

$$v(T) = \sqrt{2gH(T)/(1 - (a/S)^2)}. \tag{4}$$

The path of this particle is given by $x^2 = -2v^2(T)y/g$. The streakline is formed by the fluid particles flowing out at an exit time, t_{out} from $(t - T)$ to t. The path followed is given by

$$x = v(t_{\text{out}})(t - t_{\text{out}}), \quad y = -g(t - t_{\text{out}})^2/2 \tag{5}$$

which can be written as

$$x^2 = -2v^2(t_{\text{out}})/g. \tag{6}$$

From (5), we know that the particle located at (x, y) at time t flowed out at a time,

$$t_{\text{out}} = t - \sqrt{2(-y)/g}. \tag{7}$$

Equation (6) gives the path of a particle that left the tank at t_{out}. To obtain the equation of streakline, we will have to substitute $v(t)$, time-dependent exit speed in (6).

To find $v(t)$, we have to see how the height of the water column in the tank is decreasing. From continuity equation, we can write

$$av = -SdH/dt, \quad \text{where} \quad v(T) = \sqrt{2gH(t)/(1 - (a/S)^2)}. \tag{8}$$

We have thus,

$$-\frac{dH}{\sqrt{H}} = \sqrt{\frac{2g}{(S/a)^2 - 1}} dt. \tag{9}$$

Upon integration,

$$\sqrt{H} = \sqrt{H_0} - \sqrt{\frac{g}{2(S/a)^2 - 1}} t. \tag{10}$$

The exit speed at t_{out} is

$$v(t_{out}) = \frac{S}{a}\sqrt{\frac{2g}{(S/a)^2 - 1}}\left[\sqrt{H_0} - \sqrt{\frac{g}{2(S/a)^2 - 1}}\, t_{out}\right]. \tag{11}$$

The equation for streakline is obtained by substituting this in (6) where t_{out} is taken from (5):

$$x^2 = -y\frac{4H_0}{1 - (a/S)^2}\left[1 - \frac{\sqrt{gt^2/2} - \sqrt{(-y)}}{\sqrt{H_0\{(S/a)^2 - 1)\}}}\right]^2. \tag{12}$$

This is valid for the time up to which the tank becomes empty. What is this time?

Solution 1.4 At a time t, let h be the height of the free surface above the orifice, πy^2 its area, and σ the cross-sectional area of the orifice. Then, approximately, the velocity of water at the orifice $= \sqrt{2gh}$. If u is the uniform velocity at the free surface, then

$$\pi y^2 u = \sigma\sqrt{2gh}, \implies y^4 \propto x.$$

Thus, the required form of the vessel is $y^4 = a^3h$.

Problems of Chapter 2

Solution 2.1 Direct substitution verifies the equation of continuity. The lines of flow are given by

$$dx/(-y) = dy/x = dz/0, \quad \text{i.e.} \quad x^2 + y^2 = \text{constant}, \quad z = \text{constant}. \tag{13}$$

With velocity components, $(v_x, v_y.v_z) = (-y/r^2, x/r^2, 0)$,

$$\partial v_y/\partial x = (y^2 - x^2)/r^4 = \partial v_x/\partial y. \tag{14}$$

In fact, $v_x dx + v_y dy + v_z dz = \tan^{-1}(y/x)$. There is a velocity potential,

$$\phi = -\tan^{-1}(y/x) \tag{15}$$

and planes $y = \kappa x$ cut the streamlines orthogonally.

Solution 2.2 As seen in Fig. 2.1, AB is the equilibrium level of water, with h be the height above O. The inclinations of the tube to horizontal at A and B be α and β. Let θ be the inclination at a distance s from O. Suppose the water is displaced by a small distance x along the tube about the equilibrium. If u denotes the velocity, the continuity equation is simply. $\partial u/\partial s = 0$. The equation of motion is

$$\frac{\partial u}{\partial t} = -g \sin \theta - \frac{1}{\rho} \frac{\partial p}{\partial s}, \tag{16}$$

where $\sin \theta$ is just dy/ds. Integrating this equation,

$$s \frac{\partial u}{\partial t} = -gy - \frac{p}{\rho} + F(t). \tag{17}$$

Taking the values at the ends of the water,

$$(a + x)\frac{\partial u}{\partial t} = -g(h + x \sin \alpha) - P_{atm}/\rho + F(t),$$

$$-(b - x)\frac{\partial u}{\partial t} = -g(h - x \sin \beta) - P_{atm}/\rho + F(t). \tag{18}$$

Thus, $(a + b)\partial u/\partial t = -gx(\sin \alpha + \sin \beta)$. Realizing that $u = dx/dt$, the equation represents oscillations with the time-period, (2.49).

Solution 2.3 A detailed treatment may be found for this in Lamb's book [52].

Problems of Chapter 3

Solution 3.1 Refer to Chap. 9.

Solution 3.2 Use first law of thermodynamics, $de = -pd(1/\rho) + dq$ with $dq = Tds$ and $h = e + p/rho$ show that

$$-\frac{\nabla p}{\rho} = T\nabla s - \nabla h \tag{19}$$

Therefore, Navier-Stokes equation now takes the form of Crocco's equation

$$\frac{\partial u}{\partial t} + \nabla \left(h + \frac{1}{2} u \cdot u \right) = u \times \omega + T\nabla s \tag{20}$$

Thus, under the assumptions of steady-state ($\partial/\partial t$), constant stagnation enthalpy ($h + u^2/2 = \text{constant}$), isentropic ($\nabla s = 0$) flow becomes irrotational ($\omega = 0$).

Solution 3.3 For sink vortex the velocity is written as

$$u_\phi = \frac{\Gamma}{2\pi r} = \frac{k}{r} \tag{21}$$

Now use Bernoulli's equation to show that the surface satisfies equation of the hyperbola in the form

$$z + \frac{A}{r^2} = P_0 \tag{22}$$

where A and P_0 are the constants.

Problems of Chapter 4

Solution 4.1 For an ideal fluid, Navier-Stokes equation becomes

$$\frac{\partial u}{\partial t} + u \cdot \nabla u = -\frac{\nabla p}{\rho}$$

Using the identity $u \cdot \nabla u = \nabla(u^2/2) - u \times (\nabla \times u)$ and noting for an ideal fluid $\nabla \times u = 0$, we get

$$\frac{\partial u}{\partial t} + \nabla\left(\frac{u^2}{2}\right) + \nabla\left(\frac{p}{\rho}\right) = 0 \tag{23}$$

Now, substitute $u = \nabla\phi$ to get

$$\frac{\partial \phi}{\partial t} + (\nabla\phi)^2 + \frac{p}{\rho} = \text{constant} \tag{24}$$

Solution 4.2 For streamfunction (ψ) get the expression for potential (ϕ) as

$$\phi = -U_0 \cos\theta \left[r + \frac{R_0^2}{r} \right]$$

Now substitute this ϕ in the Eq. (24).

Solution 4.4 Inside the boundary layer, viscous and inertial terms are comparable i.e. $U_0^2/x \sim \nu U_0/\delta^2$. Therefore,

$$\nu \sim \frac{U_0 \delta^2}{x} \tag{25}$$

Now, from continuity equation the y-component of velocity scales as

$$v \sim \frac{\delta}{x} (U_0) \tag{26}$$

Now, we write the momentum equation for the y-component of velocity.

$$\frac{\partial v}{\partial t} + u \frac{\partial v}{\partial x} + v \frac{\partial v}{\partial y} = -\frac{1}{\rho} \frac{\partial p}{\partial y} + \nu \left(\frac{\partial^2 v}{\partial x^2} + \frac{\partial^2 v}{\partial y^2} \right) \tag{27}$$

Now, using the scaling $\partial/\partial x \simeq 1/x$ and $\partial/\partial x \simeq 1/\delta$ along with substituting the expressions for v and v, we can show that all viscous and inertial terms are of order δ/x or higher. Dropping these terms, we get inside the boundary layer

$$\frac{\partial p}{\partial y} \simeq 0 \tag{28}$$

Problems of Chapter 5

Solution 5.1 We seek the solution of the following form:

$$u(y, z) = k \left(\frac{y^2}{a^2} + \frac{z^2}{b^2} - 1 \right)$$

This solution satisfies the boundary condition. Substituting this in the Poisson equation we can get the expression for k.

Solution 5.2 The velocity function may be written as

$$v(x, y) = A(x^2 - 3xy^2) + B - p(x^2 + y^2)/4\mu. \tag{29}$$

This must vanish on the boundary of the equilateral triangle of sidelength a. We may determine A, B so that $v(x, y)$ vanishes at $x = a$, a part of the boundary. Thus,

$$v(x, y) = -(p/12a\mu)(x - a)(x - \sqrt{3}y + 2a)(x + \sqrt{3}y + 2a). \tag{30}$$

The flux is given by

$$\int_{\text{triangle}} v(x, y) dx dy = \frac{9\sqrt{3} \, pa^4}{20 \, \mu}. \tag{31}$$

Solution 5.3 Consider steady flow between two rotating cylinders. For solution can be readily found as

$$u_\phi(r) = \frac{1}{R_2^2 - R_1^2} \left[\left(\Omega_2 R_2^2 - \Omega_1 R_1^2 \right) r - (\Omega_2 - \Omega_1) \frac{R_1^2 R_2^2}{r} \right] \tag{32}$$

Now take the limit $R_2 \to \infty$ and $\Omega_2 = 0$ to get

$$u_\phi(r) = \frac{\Omega_1 R_1^2}{r} \tag{33}$$

Problems of Chapter 6

Solution 6.1 For 2D Stokes flow in $x - y$ plane we have

$$\nabla^2 \omega_z = 0$$

But $\omega_z = \partial v/\partial x - \partial u/\partial y$. Therefore, using definition of streamfunction (ψ) we have

$$\omega_z = \nabla^2 \psi,$$

Substituting in the vorticity equation we get the biharmonic equation for ψ in the form

$$\nabla^4 \psi = 0$$

Solution 6.3 Consider a circular stamp of radius a moving vertically downwards with velocity W onto the ink pad. The velocity components of the upper surface of the thin viscous layer are therefore, $(0, 0, -W)$ and we can express the Reynolds equation in the form,

$$\frac{\partial}{\partial x} \left(h^3 \frac{\partial p}{\partial x} \right) + \frac{\partial}{\partial y} \left(h^3 \frac{\partial p}{\partial y} \right) = -12\mu W,$$

μ being the viscous coefficient of the thin film and $U, V = 0$. Here, the thickness, h is independent of the x, y and is a function of z alone while the vertical velocity, W is independent of x, y and is a function of, however t. The Reynolds equation can be written as

$$h^3 \nabla^2 p = -12\mu W, \tag{34}$$

as the velocity components of the upper surface involve only W, (U, V) being zero. In cylindrical geometry, this equation takes the form,

$$h^3 \left[\frac{d^2 p}{dr^2} + \frac{1}{r} \frac{dp}{dr} \right] = -12 \mu W,$$

or,

$$\frac{1}{r} \frac{d}{dr} \left(r \frac{dp}{dr} \right) = -\frac{12 \mu W}{h^3},$$

which integrates to give

$$p(r) = -\frac{3 \mu W}{h^3} r^2 + A \ln r + B \tag{35}$$

Since the pressure is finite at $r = 0$, we have $A = 0$ and also since $p = 0$ at $r = a$, we have

$$p(r) = -\frac{3 \mu W}{h^3} \left(r^2 - a^2 \right) \tag{36}$$

The force resisting the motion of the rubber stamp is given by

$$\int_0^a 2\pi r \, dr \cdot p(r) = \frac{3\pi \mu W}{2h^3} a^4 \tag{37}$$

Notice the force to pull the stamp keeps increasing as h gets smaller. If we set $W = dh/dt$, then the force

$$\frac{3\pi \mu W}{2h^3} a^4 = \frac{3\mu \tau^2}{8\pi} \frac{d}{dt} \left(\frac{1}{h^4} \right),$$

where $\pi a^2 h = \tau$ is the volume occupied by the film. Thus, the motion of the stamp in direction normal to the ink pad gives rise to a force on the pad in that direction equal to

$$F = -\frac{3\pi \mu a^4}{2h^3} \frac{dh}{dt}.$$

Consequently, a constant force, F applied will be able to pull the rubber stamp away from the ink pad in time $\sim 3\pi \mu a^4 / 4F h^2$.

Problems of Chapter 7

Solution 7.1 In steady state, since the change in kinetic energy is neglected, the dissipated power will equal the work done due to pressure difference. Denote the pressure at point x by p_x. The dissipated power through all the branches is

$$P = Q_1(p_1 - p_x) + 2Q_2(p_x - p_2). \tag{38}$$

Eliminate Q_2. Note that Q_1, Q_2 are given by the Poiseuille law:

$$Q_1 = \frac{\pi}{8\mu} \frac{a_1^4(p_1 - p_x)}{x}, \quad Q_2 = \frac{\pi}{8\mu} \frac{a_2^4(p_x - p_2)}{\sqrt{d^2 + (L - x)^2}}. \tag{39}$$

Conservation of flux implies that $Q_1 = 2Q_2$. From these, p_x is obtained as a function of x. Using (38), we get P which can be minimized w.r.t. x. This condition gives

$$x = L - \frac{2a_2^4 d}{\sqrt{a_1^8 - 4a_2^8}}. \tag{40}$$

From Fig. 7.2, $\cos\theta/2 = (L - x)/\sqrt{d^2 + (L - x)^2}$, we get

$$\theta = 2\cos^{-1}[(a_2/a_1)^4]. \tag{41}$$

Solution 7.2 The net force in the positive x-direction on the volume element is seen to be

$$pA + p_0 \frac{\partial A}{\partial x}\Delta x - \left[pA + \frac{\partial(pA)}{\partial x}\Delta x\right] = -\frac{\partial}{\partial x}[(p - p_0)A]\Delta x.$$

The mass $\rho A \Delta x$ times acceleration is given by

$$\rho A \Delta x \left[\frac{\partial \rho}{\partial t} + u \frac{\partial \rho}{\partial x}\right]. \tag{42}$$

The equation of motion is

$$\rho A \left[\frac{\partial \rho}{\partial t} + u \frac{\partial \rho}{\partial x}\right] = -\frac{\partial}{\partial x}[(p - p_0)A]. \tag{43}$$

The equation of continuity is obviously

$$\frac{\partial A}{\partial t} + \frac{\partial}{\partial x}(Au) = 0. \tag{44}$$

A further relation is to be supplied between A and p, we consider (7.2). Let us linearize (43), (44), (7.2) by assuming that u, $p - p_0$, $A - A_0$ and their derivatives are small. Expanding in Taylor series about A_0, and keeping up to the first order terms, we obtain

$$\rho \frac{\partial u}{\partial t} = -\frac{\partial p}{\partial x},$$

$$\frac{\partial A}{\partial t} + A_0 \frac{\partial u}{\partial x} = 0,$$

$$p - p_0 = \frac{Yh}{2r_0 A_0}(A - A_0). \tag{45}$$

Differentiating the first equation w.r.t. x and second w.r.t. t, elimination of u gives

$$\frac{\partial^2 A}{\partial t^2} = \frac{A_0}{\rho} \frac{\partial^2 p}{\partial x^2}. \tag{46}$$

From the last equation in (45), we can express the derivative of A in terms of derivative of p, hence the wave equation:

$$\frac{1}{c^2} \frac{\partial^2 A}{\partial t^2} = \frac{\partial^2 A}{\partial x^2}, \quad c = \sqrt{Yh/2\rho r_0}. \tag{47}$$

Problems of Chapter 8

Solution 8.1 Incorporating Coriolis term corresponding to Earth's rotation, the Navier-Stokes equation is [109]

$$2\rho \boldsymbol{v} \times \boldsymbol{\omega} + \mu \nabla^2 \boldsymbol{v} - \nabla p + \mathbf{F} = 0. \tag{48}$$

The gravity vector $\mathbf{F}$ is perpendicular to ξ and η. For the simplified situation here, the ξ—and η—components of (48) for a point at geographical latitude ϕ on the northern hemisphere are:

$$2\omega \sin \phi \, v_\eta + \nu \frac{d^2 v_\xi}{d\zeta^2} = 0,$$

$$-2\omega \sin \phi \, v_\xi + \nu \frac{d^2 v_\eta}{d\zeta^2} = 0 \tag{49}$$

Since $v_\zeta = 0$ and the vertical component of the Coriolis acceleration is very small compared to gravity g, the ζ-component of (48) gives the gradient of hydrostatic pressure:

$$- dp/d\zeta = \rho g. \tag{50}$$

Equations (49) can be conveniently expressed in the complex form $V = v_\xi + i v_\eta$. Equation (49) becomes

$$d^2V/d\zeta^2 - 2i\lambda^2 V = 0, \quad \lambda^2 = \omega \sin\phi/v. \tag{51}$$

Taking care of the orientation of ζ, on integration yields $V(\zeta) = A \exp((1+i)\lambda\zeta)$. Excitation due to wind pressure determines A. Components of wind pressure which balances friction pressures $p_{\zeta\xi}$ and $p_{\zeta\eta}$ on the surface $\zeta = 0$. We introduce a complex friction pressure P by

$$P = p_{\zeta\xi} + p_{\zeta\eta} = -\mu \left(\frac{dv_\xi}{d\zeta} + i\frac{dv_\eta}{d\zeta} \right) = -\mu \frac{dV}{d\zeta}, \tag{52}$$

where v_ζ has been neglected. From the form of V for $\zeta = 0$, we have

$$- P_0 = \mu\lambda A(1+i) \tag{53}$$

which is the value of the wind pressure on the surface $\zeta = 0$. We obtain the velocities v_ξ, v_η by introducing the notation:

$$P_0 = |P_0|e^{i\delta}, \quad v_0 = -\frac{|P_0|}{\mu\lambda\sqrt{2}}, \quad \zeta = -z. \tag{54}$$

Separating the real and imaginary parts, we get

$$v_\xi = v_0 e^{-\lambda z} \cos(\lambda z + \pi/4 - \delta), \quad v_\eta = -v_0 e^{-\lambda z} \sin(\lambda z + \pi/4 - \delta). \tag{55}$$

This implies that the drift current on the surface makes an angle of $-45°$ with the wind direction. The magnitude $|v|$ reduces with depth and changes its direction in such a way that the current becomes opposite to the surface current at the depth,

$$z_{Ekman} = \frac{\pi}{\lambda} = \frac{\pi}{\sqrt{\omega\sin\phi/v}}; \tag{56}$$

$|v| = v_0 e^{-\pi}$, called Ekman's friction-depth.

Problems of Chapter 9

Solution 9.1 Include the electron pressure term in the generalized Ohm's law.

$$\mathbf{E} = -\frac{\mathbf{V} \times \mathbf{B}}{c} - \frac{\nabla p_e}{e n_e} + \frac{\mathbf{J}}{\sigma} \tag{57}$$

Substitute this expression in the magnetic induction equation and substitute $p_e = n_e T_e$ to get

$$\frac{\partial \mathbf{B}}{\partial t} = \nabla \times (\mathbf{v} \times \mathbf{B}) - \frac{\nabla n_e \times \nabla T_e}{e n_e} + \eta \nabla^2 \mathbf{B} \tag{58}$$

Note that the second term in the above equation is a source of spontaneous generation of magnetic field when the density and temperature gradients are not aligned with each other. By the term 'spontaneous' we mean that unlike the other two terms in the above equation, this term can produce magnetic field even when there is no initial magnetic field. Therefore, this term is considered as one of the candidate to explain origin of cosmic magnetic fields.

Solution 9.2 Inside the control volume the vacuum magnetic field must satisfy following equation:

$$\nabla \times \mathbf{B} = 0,$$

Since there are no currents in the control volume, the RHS of the above equation is set to zero. This also implies that the magnetic field can now be expressed as $\mathbf{B} = \nabla \psi$ where ψ is the scalar function. So, in terms of this scalar function the $\nabla \cdot \mathbf{B} = 0$ condition now become $\nabla^2 \psi = 0$.

Solution 9.3 Let us assume that minimum energy configuration of magnetic field is represented by $\mathbf{B}_{min}$. Now, we perturb this configuration by $\delta \mathbf{B}$. Since the boundary condition on the surface bounding the control volume are prescribed $\delta \mathbf{B}$ on the boundary mush vanish. Therefore, magnetic energy (E) inside the volume is given by

$$E = \frac{1}{4\pi} \int_v (\mathbf{B}_{min} + \delta \mathbf{B})^2 \, dV = \tag{59}$$

For arbitrary $\delta \mathbf{B}$ for E to be minimum energy state associate with the magnetic configuration $\mathbf{B}_{min}$ we need

$$\int_v (\delta \mathbf{B} \cdot \mathbf{B}_{min}) \, dV = 0 \tag{60}$$

Expressing perturbed magnetic field in terms of perturbed vector potential (δA) as $\delta B = \nabla \times \delta A$ and using vector identity $\nabla \cdot (A \times B) = B \cdot \nabla \times A - A \cdot \nabla \times B$, we get

$$\int_v (\delta A \cdot \nabla \times B_{min}) \, dV = - \int_v \nabla \cdot (\delta A \times B) \, dV \tag{61}$$

Using divergence theorem, RHS can be transformed into a surface integral which vanishes because perturbed magnetic field (δA) must go to zero on the surface. Therefore,

$$\int_v \delta A \cdot (\nabla \times B_{min}) = 0 \tag{62}$$

For arbitrary perturbed magnetic field, this implies

$$\nabla \times B_{min} = 0 \tag{63}$$

This proves that the minimum energy magnetic field configuration must be vacuum field configuration.

Solution 9.4 From mass conservation we have

$$V_{in}\Delta = V_{out}\delta. \tag{64}$$

Now, energy conservation can be written as $B_0^2/8\pi = \rho V_{out}^2/2$ which gives outflow velocity as

$$V_{out} = \sqrt{\frac{B_0^2}{4\pi\rho}} = V_A \tag{65}$$

Electric field inside the sheet is given by

$$E \simeq J/\sigma = \frac{c}{4\pi\sigma} (\nabla \times B) \sim \frac{cB_0}{4\pi\sigma\delta} \tag{66}$$

Electric field outside the sheet is given by

$$E \simeq -\frac{V}{c} \times B \sim \frac{V_{in}B_0}{c} \tag{67}$$

Equating the electric fields inside and outside the sheets we get

$$V_{in} \simeq \frac{c^2}{4\pi\sigma\delta} = \frac{\eta}{\delta} \tag{68}$$

where η is the magnetic diffusivity. Finally, substituting this V_{in} and $V_{out} = V_A$ in mass conservation equation we get

$$R = \frac{V_{in}}{V_{out}} = \frac{\delta}{\Delta} \simeq \sqrt{\frac{\eta}{V_A \Delta}} = \frac{1}{\sqrt{S}} \tag{69}$$

Problems of Chapter 12

Solution 12.1 In Fig. 1, the incident shock I strikes the wall at an angle α to produce a reflected shock at an angle α'. Consider a coordinate system in which the point of intersection O is at rest. The change in components of velocity normal to the shocks can be found from the equations [38]:

$$\frac{\rho_2}{\rho_1} = \left(\frac{2}{(\gamma + 1)M_1^2} + \frac{\gamma - 1}{\gamma + 1} \right)^{-1},$$

$$\frac{p_2}{p_1} = 1 + \frac{2\gamma}{\gamma + 1}(M_1^2 - 1) \tag{70}$$

where a_1 is local speed of sound; Mach number of incident shock is $M_1 = v_1/a_1$. Mach number of flow leaving the shock,

$$M_2^2 = \frac{v_2^2}{a_2^2} = \frac{1 + \frac{\gamma-1}{\gamma+1}(M_1^2 - 1)}{1 + \frac{2\gamma}{\gamma+1}(M_1^2 - 1)} \tag{71}$$

Fig. 1 Reflection of a shock

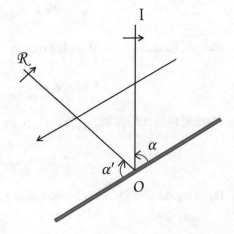

Fig. 2 Relation between incident and reflected angles is schematically shown, originally obtained by numerical solution in [91]

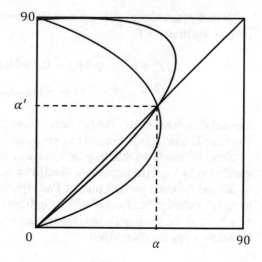

while the tangential component of velocities are unaffected. These equations have been numerically solved [91] for a certain value of the ratio, $\xi = p_1/p_2$ ($\xi = 1$, sonic case; $\xi = 0$, infinitely strong shock) (Fig. 2). Only for sufficiently small values of α ($\alpha < \alpha_0$), solutions exist. Below these extreme values, two possible reflected waves can occur.

Problems of Chapter 14

Solution 14.1 The variation of the drag coefficient with Reynolds number may be assumed to be similar to that in flow past a circular cylinder. kinematic viscosity of air is 10^{-5} m^2/s approximately. Reynolds number for the cricket ball is then approximately vd/ν, i.e. $(40 \, \text{m/s})(0.07 \, \text{m})/(10^{-5} \, \text{m}^2/\text{s}) = 280{,}000$. For the table tennis ball, it is approximately 45,000.

Solution 14.2 Written in polar coordinates in a plane, the steady motion can be expressed in terms of the radial component v_r:

$$v_r \frac{dv_r}{dt} = -\frac{1}{\rho} \frac{\partial p}{\partial r} + \nu \left(\frac{\partial^2 v_r}{\partial r^2} + \frac{1}{r} \frac{\partial v_r}{\partial r} + \frac{1}{r^2} \frac{\partial^2 v_r}{\partial \theta^2} - \frac{v_r}{r^2} \right),$$

$$-\frac{1}{r\rho} \frac{\partial p}{\partial \theta} + \frac{2\nu}{r^2} \frac{\partial v_r}{\partial \theta} = 0,$$

$$\frac{\partial v_r}{\partial r} + \frac{v_r}{r} = 0. \tag{72}$$

The last equation is satisfied generally by $r v_r = f(\theta)$.

To find f, substitute in (72), eliminate p by differentiating the equations. We get [96] an equation for f:

$$2ff' + v(f''' + 4f') = 0, \quad \text{which can be integrated},$$

$$f^2 + v(f'' + 4f) = \text{constant}. \tag{73}$$

The solution must satisfy the boundary conditions, in that the velocity vanishes at $\theta = \pm\alpha$. The exact solution can be expressed in terms of elliptic integrals.

Here, Reynolds number may be taken as mean flux over a circular cross section multiplied by length of the section divided by v. For small R, the flow will be similar to the one between parallel planes. For higher R, if the stream is converging to a sink, the velocity distribution will be uniform except near the walls at a distance $\sim 1/\sqrt{R}$. If, on the other hand, the stream is diverging from a sink, there will be backflow along the planes [96].

References

1. H.M. Antia et al., Solar Phys. **192**, 459 (2000)
2. D. Arcy Thompson, *Growth and Form*, vol. I, 2nd edn. (Cambridge University Press, 1942)
3. R.J. Arms, F.R. Hama, Phys. Fluids **8**, 553 (1965)
4. G. Auberson, S.R. Jain, A. Khare, Phys. Lett. A **267**, 293 (2000)
5. H.W. Babcock, Astrophys. J. **133**, 572 (1961)
6. H.W. Babcock, T.G. Cowling, Mon. Not. R. Astron. Soc. **113**, 357 (1953)
7. S. Babuin, M. Stammeier, E. Varga, M. Rotter, L. Skrbek, Phys. Rev. B **86**, 134515 (2012)
8. C. Barenghi, C.A. Jones, J. Fluid Mech. **197**, 551 (1988)
9. G.T. Batchelor, *An Introduction to Fluid Mechanics*, First Indian edn. (Cambridge University Press, New Delhi, 1993)
10. J.D. Bekenstein, Phys. Rev. D **7**, 2333 (1973)
11. S.T. Beliaev, Sov. Phys. JETP **7**, 289 (1958)
12. N. Berloff, A.J. Youd, Phys. Rev. Lett. **99**, 145301 (2007)
13. D. Bernard, G. Boffetta, A. Celani, G. Falkovich, Nat. Phys. **2**, 124 (2006)
14. G.P. Bewley, D.P. Lathrop, K.R. Sreenivasan, Nature **441**, 588 (2006)
15. G.P. Bewley, M.S. Paoletti, K.R. Sreenivasan, D.P. Lathrop, Proc. Natl. Acad. Sci. **105**, 13707 (2008)
16. L.Z. Bierman, Naturforsch. **52**, 65 (1950)
17. H. Bondi, Mon. Not. R. Astron. Soc. **112**, 195 (1952)
18. J. Cardy, G. Falkovich, K. Gawedzki, *Non-equilibrium Statistical Mechanics and Turbulence* (The London Mathematical Society/Cambridge University Press, Cambridge, 2008)
19. S. Chandrasekhar, *Hydrodynamic and Hydromagnetic Stability* (Courier Corporation, 2013)
20. T.G. Cowling, Mon. Not. R. Astron. Soc. **94**, 39 (1933)
21. T.G. Cowling, Magnetohydrodynamics (Crane, Russak and Company, New York, 1976)
22. T.G. Cowling, Annu. Rev. Astron. Astrophys. **19**, 115 (1981)
23. R.J. Donnelly, Annu. Rev. Fluid Mech. **25**, 325 (1993)
24. R.J. Donnelly, P.H. Roberts, Philos. Trans. R. Soc. Lond. A **271**, 41 (1971)
25. R.J. Donnelly, C.E. Swanson, J. Fluid Mech. **173**, 387 (1986)
26. J.R. Dorfman, H. van Beijeren, T. Kirkpatrick, *Contemporary Kinetic Theory of Matter* (Cambridge University Press, Cambridge, 2021)
27. J.A. Eddy, Science **192**, 1189 (1976)
28. L. Euler, Mém. Acad. Sci. Berlin **11**, 274 (1757)
29. A.P. Finne et al., Nature **424**, 1022 (2003)
30. R.P. Feynman, Phys. Rev. **91**, 1291 (1953)
31. R.P. Feynman, *Program in Low Temperature Physics*, vol. 1, ed. by C.J. Gorter (North-Holland, Amsterdam, 1955), p. 17
32. R.P. Feynman, Rev. Mod. Phys. **29**, 205 (1957)
33. U. Frisch, *Turbulence* (Cambridge University Press, New Delhi, 1999)
34. U. Frisch, S.A. Orszag, Phys. Today 43, 24 (1990)

© The Author(s), under exclusive license to Springer Nature Switzerland AG 2022
S. R. Jain et al., *A Primer on Fluid Mechanics with Applications*,
https://doi.org/10.1007/978-3-031-20487-6

35. C.S. Gardner, J.M. Greene, M.D. Kruskal, R.M. Miura, Phys. Rev. Lett. **19**, 1095 (1967)
36. H.L. Grant, R.W. Stewart, A. Moilliet, J. Fluid Mech. **12**, 241 (1962)
37. A. Griffin, T. Nikuni, E. Zaremba, *Bose Condensed Gases at Finite Temperature* (Cambridge University Press, 2009)
38. W.C. Griffith, W. Bleakney, Am. J. Phys. **22**, 597 (1954)
39. S.W. Hawking, Commun. Math. Phys. **43**, 199 (1975)
40. R. Howard, Astrophys. J. **210**, L159 (1976)
41. S.R. Jain, unpublished (2010). This was found by S. R. Jain while delivering a course on "Nonlinear plasma theory" to students of Homi Bhabha National Institute in 2010
42. S.R. Jain, *Mechanics, Waves, Thermodynamics: An Example-based Approach* (Cambridge University Press, Cambridge, 2016)
43. S.R. Jain, R. Samajdar, Rev. Mod. Phys. **89**, 045005 (2017)
44. P.L. Johnson, Phys. Rev. Lett. **124**, 104501 (2020)
45. C. Jones, C. Barenghi, J. Fluid Mech. **197**, 551 (1988)
46. T. Kirkpatrick, J.R. Dorfman, Phys. Rev. A **28**, 2576 (1983)
47. T. Kirkpatrick, J.R. Dorfman, J. Low Temp. Phys. **58**, 301 (1985)
48. T. Kirkpatrick, J.R. Dorfman, J. Low Temp. Phys. **59**, 1 (1985)
49. M. Kobayashi, M. Tsubota, Phys. Rev. Lett. **97**, 145301 (2006)
50. A.N. Kolmogorov, Dokl. Akad. Nauk SSSR **30**, 299 (1941)
51. F. Krause, K.-H. Rädler, *Mean-field Magnetohydrodynamics and Dynamo Theory* (Pergamon, London, 1980)
52. H. Lamb, *Hydrodynamics*, 6th edn. (Dover, 2009)
53. L.D. Landau, Phys. Rev. **60**, 356 (1941)
54. L.D. Landau, E.M. Lifshitz, *Fluid Mechanics*, 2nd edn. (Elsevier, Oxford, 1987)
55. E. Lauga, T.R. Powers, Rep. Progr. Phys. **72**, 096601 (2009)
56. D. Layzer, Astrophys. J. **141**, 837 (1965)
57. D. Layzer, R. Rosner, H.T. Doyle, Astrophys. J. **229**, 1126 (1979)
58. N.R. Lebovitz, Astroph. J. **134**, 500 (1961)
59. N.R. Lebovitz, S. Chandrasekhar, Astrophys. J. **135**, 238 (1962)
60. R.B. Leighton, Astrophys. J. **140**, 1547 (1964)
61. R.B. Leighton, Astrophys. J. **156**, 1 (1969)
62. R. Liboff, Quantum mechanics, in *Introductory Quantum Mechanics* (Pearson Education India, 2003)
63. H.W. Liepmann, A. Roshko, *Elements of Gas Dynamics* (Courier Corporation, 2001)
64. M.J. Lighthill, *Mathematical Biofluiddynamics* (SIAM, Philadelphia, 1975)
65. M.J.M. Lighthill, *Waves in Fluids* (Cambridge University Press, Cambridge, 1978)
66. W. Livingston, J. Harvey, Solar Phys. **10**, 294 (1969)
67. F. London, Nature **141**, 643 (1938)
68. F. London, Phys. Rev. **54**, 947 (1938)
69. E.N. Lorenz, *The Essence of Chaos* (UCL Press, 1993)
70. L.P. Kadanoff, Phys. Today **54**, 34 (2001)
71. L.P. Kadanoff, A. Libchaber, E. Moses, G. Zocchi, La Recherche **22**, 628 (1991)
72. L. Madeira, M.A. Caracanhas, F.E.A. dos Santos, V.S. Bagnato, Annu. Rev. Condens. Matter Phys. **11**, 37 (2020)
73. M. Margules, Wien. Ber. (2nd ser.) **83**, 588 (1881)
74. J. Maurer, P. Tabeling, Europhys. Lett. **43**, 29 (1998)
75. D.A. McDonald, *Blood Flow in Arteries*, 2nd edn. (Edward Arnold, London, 1974)
76. L. Mestel, Star formation and galactic magnetic field. Vistas Astron. **3**, 296 (1960)
77. L. Mestel, Mon. Not. R. Astron. Soc. **138**, 359 (1968)
78. C.W. Misner, K. Thorne, J.A. Wheeler, *Gravitation* (Freeman, San Francisco, 1973)
79. H. Moffat, Fluid Dynam. Trans. **8**, 99 (1976)
80. T. Mouschovias, Astrophys. J. **192**, 37 (1974)
81. S.K. Nemirovskii, Low Temp. Phys. **45**, 000000-841 (2019)

82. I. Newton, *Mathematical Principles*, ed. by F. Cajori (University of California Press, Berkeley, 1946), p. 385 ff.
83. H. Oertel (ed.), *Prandtl-Essentials of fluid mechanics*, 3rd edn. (Springer, Heidelberg, 2010)
84. L. Onsager, Nuovo Cim. **6**, 279 (1949)
85. M.S. Paoletti, D.P. Lathrop, Annu. Rev. Condens. Matter Phys. **2011**(2), 213 (2010)
86. E.N. Parker, Astrophys. J. **122**, 293 (1955)
87. E.N. Parker, Astrophys. J. **145**, 811 (1966)
88. J.H. Piddington, Solar Phys. **31**, 229 (1973)
89. L.P. Pitaevskii, Sov. Phys. JETP **13**, 451 (1961)
90. F. Plunian, T. Alboussière, J. Fluid Mech. **A 66**, 941 (2022)
91. H. Polachek, R. Seeger, Phys. Rev. **84**, 922 (1951)
92. S.K. Prasad, D.B. Jess, J.A. Klimchuk, D. Banerjee, Astrophys. J. **834**, 103 (2017)
93. E.R. Priest, *Solar Flare Magnetohydrodynamics* (Gordon and Breach, Chicago, 1981)
94. E.M. Purcell, Am. J. Phys. **45**, 096601 (1977)
95. M. Rabaud, F. Moisy, Phys. Rev. Lett. **110**, 214503 (2013)
96. A.S. Ramsey, *A Treatise on Hydromechanics*, Part II, First Indian edn. (CBS Publishers, New Delhi, 1988)
97. A.M. Reed, J.H. Milgram, Annu. Rev. Fluid Mech. **34**, 469 (2002)
98. L.F. Richardson, *Weather Prediction by Numerical Process* (Cambridge University Press, Cambridge, 1922)
99. P.H. Roberts, R.J. Donnelly, Phys. Lett. A **31**, 137 (1970)
100. S.I. Rubinow *Introduction to Mathematical Biology* (Dover, New York, 2002)
101. K.W. Schwarz, Phys. Rev. B **38**, 2398 (1988)
102. L.I. Sedov, A.G. Volkovets, *Similarity and Dimensional Methods in Mechanics* (CRC Press, 2018)
103. A.L. Selby, Lond. Edinb. Dublin Philos. Mag. J. Sci. **29**, 113 (1890)
104. L. Skrbek, K.R. Sreenivasan, Phys. Fluids **24**, 011301 (2012)
105. L. Skrbek, W.F. Vinen, *Program in Low Temperature Physics*, vol. 16, ed. by M. Tsubota, W.P. Halperin (Elsevier, Amsterdam), pp. 195–246
106. E.A. Spiegel, G. Veronis, Astrophys. J. **131**, 442 (1960)
107. L. Spitzer, *Physics of Fully Ionized Gases*, 2nd edn. (Dover, 2013)
108. L. Spitzer, L. Mestel, Star formation in magnetic dust clouds. Mon. Not. R. Astron. Soc. **116**, 503 (1956)
109. A. Sommerfeld, *Mechanics of Deformable Bodies* (Levat Books, Kolkata, 2006)
110. M. Steenbeck, F. Krause, Astron. Nachr. **291**, 49 (1969)
111. J.O. Stenflo, Astron. Astrophys. Rev. **21**, 1 (2013)
112. G.G. Stokes, *Mathematical and Physical Papers*, vols. 1 and 5 (Cambridge University Press, Cambridge, 1880 and 1905)
113. C. Swanson, R.J. Donnelly, Phys. Rev. Lett. **67**, 1578 (1991)
114. G.I. Taylor, Philos. Trans. R. Soc. Lond. A **223**, 289 (1923)
115. G.I. Taylor, Proc. R. Soc. Lond. **201**, 159 (1950)
116. L. Tisza, Nature **141**, 913 (1938)
117. J.S. Trefil, *Introduction to the Physics of Fluids and Solids* (Courier Corporation, 2012)
118. W. Unruh, Phys. Rev. D **14**, 4 (1976)
119. W. Unruh, Phys. Rev. Lett. **46**, 21 (1981)
120. W.F. Vinen, Proc. R. Soc. Lond. A **240**, 114 (1957)
121. W.F. Vinen, Proc. R. Soc. Lond. A **240**, 128 (1957)
122. W.F. Vinen, Proc. R. Soc. Lond. **242**, 493 (1957)
123. W.F. Vinen, Phys. Rev. B **64**, 134520 (2001)
124. W.F. Vinen, M. Tsubota, A. Mitani, Phys. Rev. Lett. **91**, 135301 (2003)
125. J. von Neumann, J. Res. Nat. Bur. Stand. **12**, 36 (1951)
126. R.M. Wald, *General Relativity* (Overseas Press (India) Pvt Ltd, New Delhi, 2007)
127. C. Walén, *On the Vibratory Rotation of the Sun* (University of Stockholm, 1949)
128. P.M. Walmsley, A.I. Golov, Phys. Rev. Lett. **100**, 245301 (2008)

129. C.N. Yang, Rev. Mod. Phys. **34**, 694 (1962)
130. E.J. Yarmchuk, M.J.V. Gordon, R.E. Packard, Phys. Rev. Lett. **43**, 214 (1979)
131. T. Young, Philos. Trans. R. Soc. Lond. **98**, 164 (1808)
132. V.E. Zakharov, V. Lvov, G. Falkovich, *Kolmogorov Spectra of Turbulence* (Springer, Berlin, 1992)
133. E. Zaremba, T. Nikuni, A. Griffin, J. Low Temp. Phys. **116**, 277 (1999)
134. S. Zuccher, M. Caliari, C.F. Barenghi, Phys. Fluids **24**, 125108 (2012)

Index

Printed in the United States
by Baker & Taylor Publisher Services